Phytochemicals of Nutraceutical Importance

Phytochemicals of Nutraceutical Importance

Editors

Vivek Karwande, Ravindra Borade

scitus academics

Phytochemicals of Nutraceutical Importance

Edited by **Vivek Karwande, Ravindra Borade**

Printed in 2017

ISBN: 978-1-68117-084-8

Library of Congress Control Number: 2015935482

SCITUS Academics LLC,
616, Corporate Way, Suite 2, 4766,
Valley Cottage, NY 10989

www.scitusacademics.com

Notice

Contents

Preface

Phytochemicals of nutraceuticals importance are bioactive constituents that sustain or promote health and occur at the intersection of food and pharmaceutical industries. Such substances may range from isolated nutrients, dietary supplements and specific diets to genetically engineered designer foods, herbal products, processed foods and beverages. Phytochemicals are broadly described as phytoestrogens, terpenoids, carotenoids, limonoids, phytosterols, glucosinolates, polyphenols, flavonoids, isoflavonoids and anthocyanidins. They have tremendous impact on the health care system and may provide medical health benefits including the prevention and/ or treatment of diseases and physiological disorders. Majority of foods, such as whole grains, beans, fruits, vegetables and herbs contain phytochemicals of nutraceutical importance. These phytochemicals, either alone and/or in combination, have tremendous therapeutic potential in curing various ailments. The respective health benefits are based on science and ethics, for health claims, functional foods and presence of certain phytochemicals. They play specific pharmacological effects in human health as anti-inflammatory, anti-allergic, antioxidants, antibacterial, antifungal, antispasmodic, chemopreventive, hepato-protective, hypolipidemic, neuroprotective, hypotensive, antiaging, diabetes, osteoporosis, DNA damage, cancer and heart diseases, induce apoptosis, diuretic, CNS stimulant, analgesic, protects from UVB-induced carcinogenesis, immuno-modulator and carminative.

Editor

Chapter 1

Phytochemical and Biological Activities of Four Wild Medicinal Plants

Anwar Ali Shad[1], Shabir Ahmad[2], Riaz Ullah[3], Naser M. AbdEl-Salam[4], H. Fouad[4,5], Najeeb Ur Rehman[6], Hidayat Hussain[6], and Wajid Saeed[1]

[1]Agricultural Chemistry Department, the University of Agriculture, Peshawar, Khyber Pakhtunkhwa 25000, Pakistan

[2]Department of Chemistry, Islamia College Peshawar, Khyber Pakhtunkhwa 25000, Pakistan

[3]Department of Chemistry, Government College Ara Khel, FR Kohat, Khyber Pakhtunkhwa 26000, Pakistan

[4]Riyadh Community College, King Saud University, Riyadh 11437, Saudi Arabia

[5]Department of Biomedical Engineering, Faculty of Engineering, Helwan University, Helwan 11790, Egypt

[6]UoN Chair of Oman's Medicinal Plants and Marine Natural Products, University of Nizwa, Birkat Al Mauz, 616 Nizwa, Oman

ABSTRACT

The fruits of four wild plants, namely, Capparis decidua, Ficus carica, Syzygium cumini, and Ziziphus jujuba, are separately used as traditional dietary and remedial agents in remote areas of Khyber Pakhtunkhwa, Pakistan. The results of our study on these four plants revealed that the examined fruits were a valuable source of nutraceuticals and exhibited good level of antimicrobial activity. The fruits of these four investigated plants are promising source of polyphenols, flavonoids, alkaloids, terpenoids, and saponins. These four plants' fruits are good sources of iron, zinc, copper, manganese, selenium, and chromium. It was also observed that these fruits are potential source of antioxidant agent and the possible reason could be that these samples had good amount of phytochemicals. Hence, the proper propagation, conservation, and chemical investigation are recommended so that these fruits should be incorporated for the eradication of food and health related problems.

INTRODUCTION

The wild plants play significant role in the suppression of dietary and pathogen related ailments of native people since long before recorded history. Latest research investigation observed that the bioactive and antioxidant potentials of these plants are attributed to the presence of polyphenols, flavonoids, lignins, alkaloids, terpenoids, carotenoids, vitamins, and so forth [1–3]. They help in maintaining the nutritional quality and shelf life of foods by inhibiting lipid oxidation, minimizing rancidity, and removing toxic oxidative products [4–7]. Similarly, phenolic compounds play important role in antioxidant activity and resistance against pests and other species dissemination.

The emergent population coupled with poverty and natural disasters interrogates how the native people of Khyber Pakhtunkhwa are able to cope with the challenge of easy access to food and medicine. Despite the government's mammoth expenses on the livelihood of common people, the provision of balanced food and modern healthcare to rural people is still a far-reaching goal [8]. Hence, it is recommended that researchers should resort to forms of nutraceuticals mainly in the native plant species to overcome the constraints of human necessities. Phytochemicals and minerals ingredients are necessary for virtually all reactions to occur in the body [9–11]. While each has its own unique properties, they work synergistically to ensure reactions in the body occur appropriately [12].

Southern parts of Khyber Pakhtunkhwa are blessed with plenty of wild plants which have very significant role in the daily life of rural people. The main areas of this region include Kohat, Karak, Bannu, and Dera Ismail Khan. In this regard, a project was designed to investigate four different food plants (Ficus carica, Syzygium cumini,Capparis decidua, and Ziziphus jujube) that are separately used in cuisines either in fresh form or in semicooked form. It was also observed that some people of the subject areas use them for the eradication of various diseases. The botanical description of each examined plant is described briefly as follows.

Ficus carica L. is important species of Moraceae family which is locally known as Inzar in Pushto. It is widely grown throughout the temperate world, both for its fruit and as an ornamental. The fruits and leaves parts of this plant exhibited significant level of antioxidant and antimicrobial activity [13, 14]. The fruit has a laxative effect and the daily uses of dried fig enhanced the antioxidant capacity in plasma [15]. Similarly, Jamun (Syzygium cumini L. Skeels) belongs to Myrtaceae family and is considered the richest nutritional source. It contains flavonoids, tannins, triterpenoids, carotenoids, and sitosterols. The extracts of this plant exhibited various activities including cytotoxic, anti-inflammatory, anticancer, and antidiabetic activities [16].

The fruit of Capparis decidua Forssk. Edgew. (Capparidaceae) normally known as Khair is a wonderful source of alkaloids, terpenoids, glycosides, and some fatty acids. The plant has significant hypercholesterolemic, anti-inflammatory, analgesic, antidiabetic, antimicrobial, antiplaque, antihypertensive, anthelmintic, and purgative potential [17–21]. Likewise, Ber (Ziziphus jujuba Mill.) of family Rhamnaceae widely grows in southern parts of Khyber Pakhtunkhwa province of Pakistan. The fruits and seeds are used for antifungal, antibacterial, antiulcer, anti-inflammatory, sedative, antispastic, antifertility/contraception, hypotensive, and wound healing properties [22].

MATERIALS AND METHODS

All the chemical studies including phytochemical screening, antimicrobial activity, antioxidant activity, and polyphenolic content of medicinal plants were conducted at Agricultural Chemistry Department, University of Agriculture, Peshawar, and Pharmacy Department, Peshawar University.

Plant Collection and Identification

The fresh fruit samples (Ziziphus jujuba, Capparis decidua, Ficus carica, and Syzygium cumini) were collected from different locations of Kohat, Karak, Bannu, and D. I. Khan-Khyber Pakhtunkhwa in 2008–2010. Each fruit along with aerial part of the fruit plant was identified by Mr. Shahid Farooq (Plant Taxonomist, PCSIR, Peshawar). A set of voucher specimens of each plant was deposited in the herbarium of PCSIR, Peshawar.

Sampling

About 3–5 kg samples of each plant fruit material were collected from their natural habitat from 4-5 different sites. About 1-2 kg representative sample of each plant was thoroughly mixed and immediately washed of dust, impurities, and other adhering materials using tap water. The fruit sample was then dried at room temperature under shade for 4-5 days or until the fruits were completely dried. The dried samples were first ground into small pieces by use of mortar and pestle and then into fine powder by use of electrical grinder. These powdered samples were sealed in plastic bags and stored in refrigerator until they were analyzed.

Extraction

The powder plants samples were macerated in aqueous methanol for primary extraction. The filtrates were then evaporated using rotary evaporator.

Phytochemical Screening

Phytochemical screening of the subject samples for the qualitative identification of tannins, saponins, terpenoids, steroids, and flavonoids was done by various analytical techniques. Standard procedure was used for screening of the constituents as described by [23–25].

Determination of Tannins

About 0.5 g dry powder of each sample was boiled in 20 mL of water in a test tube and then filtered. Few drops of 0.1% ferric chloride solution were added and noticed for brownish green or blue-black coloration [24].

Determination of Saponins

About 2 g of the powder sample was boiled in 20 mL of distilled water on a water bath and filtered. 10 mL filtrate was mixed with 5 mL of distilled water and shaken vigorously for a stable persistent froth. Then, the froth was mixed with 3 drops of olive oil and shaken vigorously, till emulsion formed [23].

Determination of Flavonoids

Approximately 5 mL of the dilute ammonia solution was added to a portion of the aqueous filtrate of each plant extract followed by addition of concentrated H_2SO_4. A yellow coloration was observed in each extract indicating the presence of flavonoids. The yellow coloration disappeared on standing [24].

Determination of Terpenoids

About 5 mL of each plant extract was mixed with 2 mL of chloroform and after mixing about 3 mL concentrated H_2SO_4 was carefully added to form a layer. A reddish brown coloration of the inner face will be formed indicating the presence of terpenoids [25].

Determination of Steroids

About 20 g of each plant sample was soaked in ethanol in a conical flask. The sample containing ethanol was warmly heated in water bath and then filtered. In this process the ethanol fraction was extracted and then the alcohol portion was evaporated from the filtrate by placing it in water bath in a large size watch glass. The solid sample left behind was dissolved in chloroform. Now acetic anhydride 4.5 mL and sulfuric acid 0.5 mL were added. During the experiment the color of the extract changed from violet to green, which was indicating the presence of steroids [23].

Total Polyphenols Determination

Total polyphenolic content of the crude fractions of subject plants was evaluated through the FCR method in which the gallic acid was used as standard. 100 µl of each plant sample was taken and subsequently 900 µl of distilled water was added to the samples, after which 0.5 mL of FCR was added to the mixture. 20% of the Na_2CO_3solution was made. From this solution 1.5 mL was taken and added into the above mixture (sample + water + FCR). The whole mixture was placed in volumetric flask on water bath and heated for about 2 hours. After the cooling of the mixture, UV-vis spectrophotometer was employed to measure the absorbance at the wavelength of 720 nm. Total polyphenolic contents in samples were assessed by using gallic acid equivalent (GAE) as standard [26].

Total Flavonoid Contents Determination

Total flavonoid content of the crude fraction of each of the samples was evaluated through the calorimetric method in which Quercetin was taken as standard. One mL of each sample was taken and 4 mL of distilled water was added. To this mixture sodium nitrate and $Alcl_3$ were added at the concentration mixture of 300 µl. The total volume was then incubated for 5 minutes followed by the addition of 2 mL of NaOH; the volume was increased up to 10 mL. UV-vis spectrophotometer was used for absorbing the wavelength set at 510 nm and total flavones were then expressed as Quercetin equivalents in mg/g of dry samples.

Antioxidant Activity

DPPH Radical Scavenging Assay

The powdered material of the subject plants was extracted with methanol for 36–48 h. The solvent after extraction of each sample was evaporated at low temperature under reduced pressure in rotary vacuum evaporator to obtain crude extracts.

Antioxidant activity was determined by using the 2,2-diphenyl-1-picrylhydrazyl (DPPH) radical scavenging assay as described by [27–30]. DPPH (1 mL, 0.3 mM) solution in ethanol was added to 2.5 mL (100 mg mL^{-1}) of methanol extract and mixture was allowed to react for 30 min at room temperature and absorbance was recorded using UV visible spectrophotometer at 518 nm. HPLC grade Rutin hydrate (95%, 1 mM) was used as standard, while absolute ethanol was used as negative control. The experiment was repeated in triplicate, using fresh plant sampled each time, and the average absorbance values were converted to percentage antioxidant activity.

Antibacterial and Antifungal Activities

Bacterial and fungal strains tests were performed on three bacteria and two fungi reference strains. Bacterial strains were Escherichia coli, Bacillus subtilis, and Staphylococcus aureus while fungal strains includeTrichophyton longifusus (clinical isolate) and Candida albicans. They were maintained on agar slant at 48°C. The strains were activated at 378°C for 24 hr on nutrient agar or Sabouraud glucose agar (SGA), respectively, for bacteria and for fungi, prior to any screening [31].

Minerals Assay

Dried and ground leaves samples were wet digested for minerals as stated by [32] in 1 : 1 mixture of HNO_3-$HClO_4$by gradually increasing the temperature up to 300°C. Trace elements such as iron (Fe), copper (Cu), zinc (Zn), manganese (Mn), and selenium (Se) were determined using double beam atomic absorption spectrophotometer Perkins Elmer Model 2380.

RESULTS AND DISCUSSION

In the present study, four wild plants found in southern parts of Khyber Pakhtunkhwa were investigated for their dietary and pharmacological potential. These plants (Capparis decidua, Ficus carica, Syzygium cumini, andZiziphus jujuba) were investigated for secondary metabolites using phytochemical screening, biochemical compounds and mineral content, antimicrobial potential, and antioxidant activity. In the present study, selected ethnobotanical information is displayed in Table 1. This information was obtained during the collection time from the native populace. The study of literature also confirmed its ethnobotanical potential [18–20]. These plants are separately used by local dwellers. Although various parts of the examined plants are used for different purposes, the fruit parts are very popular among the people. The fruit part of Capparis decidua plant is known as karir or kary which is edible and also used in various forms of processed food like pickles and making sweet food products. The flowers are used as a vegetable and commonly consumed with loaf and meat. The Ficus caricalocally known as Inzar is widely distributed in different parts of Khyber Pakhtunkhwa. Fruit and latex are routinely used for dietary and medicinal purposes. The fruit part is used in both fresh and dried forms and it is also used in making sweet food products like cake and bakery making [33, 34].

Table 1: Ethnobotanical information of selected species of subject area used as food and medicine

S. number	Botanical name	Family	Local name	Life form	Parts used	Mode/form of consump tion	Perceived traditional ethno botanical use
1	Capparis decidua Edgew	Capparidaceae	Karir	Shrub	Flower and fruit	Raw, dried, semicooked, and used in making pickles	Dietary and medicinal
2	Ficus carica Forssk	Moraceae	Inzar (Injeer)	Shrub/ small tree	Milky latex and fruits	Raw, dried, decoction, eaten fresh, and used in salad, jam, marmalade, and bakery products	Dietary and medicinal

3	Syzygium cumini L.	Myrtaceae	Jaman	Tree	Fruits, seeds, leaves, and bark	Raw and extracts and used in salad, jam, jellies, refreshing drinks, and other sweet food products	Dietary and medicinal (diabetes)
4	Ziziphus jujuba Mill.	Rhamnaceae	Baira	Tree	Leaves, bark, and fruits	Raw, decoction, and dried	Dietary and medicinal

Syzygium cumini locally known as Jaman is very famous for its dietary and medicinal values. The leaves, fruit, seeds, and bark are very effective against anti-inflammatory activity, asthma, diarrhea, fever, diabetes, and other common ailments [35]. The fruits are consumed in different parts of Pakistan as fresh raw fruit as such or in salad or processed in jams, squash, and so forth. Similarly the Ziziphus jujuba plant is commonly known as baira and it is one of the widely grown plants of southern parts of Khyber Pakhtunkhwa. The fruit is liked by people of all ages and used both in dried and in fresh forms. It is also used in sweet food products making and used for the eradication of bronchitis, diarrhoea, dysentery, cold, and coughs.

Phytochemicals are natural bioactive compound widely distributed in plants, animals, microbes, and other forms of life. The principal examples of natural products including alkaloids, terpenoids, steroids, polyphenols, and flavonoids have rational uses and are found in varying amounts in different species. Their presence in the food chains has significant role which is to work with nutrients and dietary fibre to protect against disease [36–38]. Nevertheless, their presence in foods has traditionally been regarded as antinutritional factor [39] and in some cases has limited their use due to their bitter taste [40, 41].

In the present study, the preliminary phytochemical screening of the subject plants reveals that they are good sources of natural products. The results concluded that the studied plants contained appreciable amount of tannins, flavonoids, steroids, alkaloids, and saponins; however terpenoids content was almost negligible (Table2). The results regarding flavonoids and tannins were very promising in the investigated plants. The preliminary phytochemical screening of Syzygium cumini and Ziziphus jujuba showed promising results for the presence of saponins, tannins, terpenoids, and flavonoids while the examined plants revealed the least level of steroids concentration in the crude aqueous methanolic extract. Similarly, Borhade [42] reported that flavonoids, tannins, steroids, and saponins are the main constituents in the Syzygium cumini plant. It is confirmed from the literature that larger amount of flavonoids might be responsible for their healing effect against pathogenic microorganisms [43]. Similarly, saponins concentration is appreciable in Capparis decidua and Ficus carica and has the property of binding with cholesterol, bitterness, and hemolytic activity in aqueous solution [44].

Table 2: Phytochemical screening of methanolic extract of investigated samples

Names	Capparis decidua	Ficus carica	Syzygium cumini	Ziziphus jujuba
Tannins	++	++	++	++
Flavonoids	++	++	++	++
Alkaloids	++	++	++	++
Phenols	++	++	++	++
Saponins	–	++	++	++
Steroids	++	++	++	++

++: present and –: absent.

In the present study all the four subject plants were investigated for total polyphenols and flavonoids content and their results are oscillated in Table 3. Phenols are the key component of many biocompounds like salicylic acid (aspirin) and recent literature studies reported that thousands of novel and variant phenolic compounds were identified and discovered in different biological sources. They are the key unit in all the foods we normally consume. The primary role of their presence in our diets is acting as antioxidant, antiseptic, and anti-inflammatory and for the eradication of other different human aliments. They also have nuisance effect when applied directly to the skin.

Table 3: The phytochemical constituents (flavonoid extract and total phenolic content) of selected plants form the studied area

Samples/extracts	Total phenolic contents* ± SD		Total flavonoid** Content ± SD	
	Hexane extract	Me.OH extract	Hexane extract	Me.OH extract
Capparis decidua Edgew	11.15 ± 4.32	23.2 ± 3.15	26.60 ± 12.28	43.46 ± 12.52
Ficus carica Forssk	5.95 ± 2.47	16.95 ± 3.23	26.95 ± 7.17	36.60 ± 12.28
Syzygium cumini L.	4.2 ± 3.28	31.2 ± 3.15	13.2 ± 3.15	25.2 ± 3.15
Ziziphus jujube Mill	5.78 ± 2.17	21.74 ± 4.35	3.25 ± 1.35	15.66 ± 3.87

*mg GAE/g of extract and **mg Quercetin equivalent/g of extract.

Flavonoids are hydroxylated phenolic substances, responsible for their therapeutic potency against wide array of microorganisms, probably due to their ability to complex with extracellular and soluble proteins and to complex with the bacterial cell wall. Flavonoids and other phenolic compounds are potent water soluble antioxidants and free radical scavengers, which prevent oxidative cell damage and have strong anticancer activity [45, 46].

The aim of this test was to evaluate the absorbing capacity of subject plants in two different fractions (hexane and methanol) by using Folin-Ciocalteu reagent. Gallic acid was used as standard to evaluate the results obtained. The results showed total phenolic content and total flavonoid contents were found in promising amount in all the studied plants ranging from 4 to 31 mg GAE/g and from 3 to 43 mg Quercetin/g, respectively. The maximum content of total phenolic compound was observed in methanolic fraction of Syzygium cumini (31.2 ± 3.15 mg GAE/g) followed by Capparis decidua (23.2 ± 3.15 mg GAE/g) and low level of the content was found in hexane fraction of Syzygium cumini (4.2 ± 3.28 mg GAE/g). Similarly, the flavonoid content of the Capparis decidua was maximum in methanolic extract (43.46 ± 12.52 mg Quercetin/g) and low level of the content was found in hexane fraction of Ziziphus jujuba (3.25 ± 1.35 mg Quercetin/g).

Research has confirmed that subject plants possess total phenolic and flavonoids contents which are of enormous benefits for mankind, thus ratifying our findings. The antioxidant support we get from investigated plants is largely due to their phenolic compounds that have been shown to help protect us against unwanted oxygen damage to our cells, blood vessels, and organ systems [47]. Antioxidant compounds in food play an important role as a health protecting factor. Scientific evidence suggests that antioxidants reduce the risk for chronic diseases including cancer and heart disease. It is increasingly being realized that many of today's diseases are due to the "oxidative stress" that results from an imbalance between formation and neutralization of prooxidants. Oxidative stress is initiated by free radicals; hence, all human beings protect themselves against free radical damage somehow by antioxidant supplements, which are vital to combat oxidative damage. It is also called free radical scavenging activity because of its ability to scavenge free radicals. Free radical generation is directly related with oxidation in foods and biological systems. Therefore, the search for potential free radical scavenging agent is imperative and indispensable.

Table 4 displays the percent antioxidant activity of studied plants using 2,2-diphenyl-1-picrylhydrazyl (DPPH) radical. In this study, three different concentrations were selected (0.25, 0.50, and 1.00 mg). It was observed that highest antioxidant activity was observed in 1.00 mg content of Syzygium cumini (63.3%) followed by Ziziphus jujuba (47.3%), Capparis decidua (15.5%), and Ficus carica (10.9%), respectively. However, slightly different trends were observed in analyzing the 0.25 and 0.5 mg content of the examined plants. In this case, Ficus carica exhibited maximum antioxidant activity at 0.5 mg level (27.3%) followed by 0.25 mg level (19.5%).

Table 4: Antioxidant activity (%) of the studied plants found from various regions of KPK

Plant species	Concentration (mg)		
	0.25	0.50	1.00
Capparis decidua	9.1	12.3	15.5
Ficus carica Forssk	27.3	19.1	10.9
Syzygium cumini L.	35.0	49.2	63.3
Ziziphus jujuba	34.9	41.1	47.3

In the present research study, methanolic crude extracts of Capparis decidua, Ficus carica, Syzygium cumini, andZiziphus jujuba were evaluated for antimicrobial potential (Table 5). They were investigated against three different bacterial strains (E. coli, P. aeruginosa, and S. typhi) and two different fungal strands includingTrichophyton longifusus and Candida albicans. Minimum inhibitory concentrations of the effective extracts were estimated by agar dilution method. The results were compared with that of standard drug.

Table 5: Antimicrobial (antifungal and antibacterial) assay at 5 mg/mL of methanolic crude extracts of selected plants (average value ± SD, µg/mL)

Fractions	EC	PA	ST	TL	CA
Capparis decidua	17.7 ± 6.58	11.16 ± 2.05	0.96 ± 0.1	0.82 ± 0.28	1.11 ± 0.50
Ficus carica	0.74 ± 0.15	0.67 ± 0.24	0.12 ± 0.03	1.17 ± 0.30	0.5 ± 0.2
Syzygium cumini	0.73 ± 0.20	10.0 ± 1.64	0.54 ± 0.23	0.46 ± 0.15	0.46 ± 0.01
Ziziphus jujuba	0.60 ± 0.1	3.40 ± 0.49	0.93 ± 0.16	0.85 ± 0.07	—
Cipro/Ampho.B	0.06	1.5	0.25	0.1	0.5

EC: E. coli; PA: P. aeruginosa; ST: S. typhi, TL: Trichophyton longifusus; CA: Candida albicans. Cipro and Ampho.B (standard drug), SD: standard deviation.

Similarly, P. aeruginosa is a frequent cause of nosocomial infections such as pneumonia, urinary tract infections (UTIs), and bacteremia [48]. Salmonella typhi is the main food poisoning agent and typhoid fever that is a major public health problem worldwide.

Likewise, the pathogenic fungi Trichophyton longifusus are responsible for the infections of dermatophytosis in the hair, skin, and nails of humans due to their ability to utilize keratin as a source of nourishment. Another fungal strain Candida albicans is responsible for the candidiasis which affects various body parts including ear, mouth, throat, skin, scalp, fingers, toes, nails, bronchi, lungs, and intestine gastrointestinal tract. Factors predisposing people to candidiasis include AIDS, burn patients, young individual, pregnancy, oral birth control, high fruit diets, steroids, antibiotic therapy, immune suppressants, cancer treatments, heart surgery, genetic deficiency, and endocrine deficiency diabetes.

Results of the study divulged that Capparis decidua, Ficus carica, Syzygium cumini, and Ziziphus jujuba had potent antibacterial agent against E. coli, S. typhi, and P. aeruginosa. It was observed that crude methanolic fraction of Ficus carica, Syzygium cumini, and Ziziphus jujuba exhibited good level of activity against all investigated strains. Similarly, the Ziziphus jujuba is effective against E. coli showing MIC value of 0.06±0.1 and is least effective against P. aeruginosa. E. coli is one of the most frequent agents of many bacterial infections. The major diseases caused by this strain are cholecystitis, bacteremia, cholangitis, urinary tract infection, diarrhoea, neonatal meningitis, and pneumonia [49]. Similarly, Ficus carica and Syzygium cumini exhibited activity against both bacterial and fungal strains. The possible reasons of this activity might be due to the presence of phytochemicals. The literature studies reveal that these natural products have significant role in the biological and pharmacological activity.

Likewise, the investigated plants were also analysed for the elemental composition. Mean values of the data reported in Table 6 showed that manganese and iron are found in maximum concentration followed by zinc whereas Cr is found in minimum level. Manganese showed a concentration ranging from 0.53 to 12 mg 100^{-1} g. It has been reported previously that chromium nutrition leads to decreased

9. J. A. Vinson, Y. Hao, X. Su, and L. Zubik, "Phenol antioxidant quantity and quality in foods: vegetables,"Journal of Agricultural and Food Chemistry, vol. 46, no. 9, pp. 3630–3634, 1998.
10. T. K. Pilvi, S. Harala, R. Korpela, and E. M. Mervaala, "Effects of high-calcium diets with different whey proteins on weight loss and weight regain in high-fat-fed C57BL/6J mice," British Journal of Nutrition, vol. 102, no. 3, pp. 337–341, 2009.
11. P. Dhan, C. Gupta, and G. Sharma, "Importance of phytochemicals in nutraceuticals," Journal of Chinese Medicine Research and Development, vol. 1, no. 3, pp. 70–78, 2012.
12. D. L. Watts, "Nutrient interrelationships: minerals, vitamins, endocrines," Journal of Orthomolecular Medicine, vol. 5, no. 1, pp. 11–19, 1990.
13. M. E. Kislev, A. Hartmann, and O. Bar-Yosef, "Early domesticated fig in the Jordan Valley," Science, vol. 312, no. 5778, pp. 1372–1374, 2006
14. M.-R. Jeong, H.-Y. Kim, and J.-D. Cha, "Antimicrobial activity of methanol extract from Ficus carica leaves against oral bacteria," Journal of Bacteriology and Virology, vol. 39, no. 2, pp. 97–102, 2009
15. R. Veberic, M. Colaric, and F. Stampar, "Phenolic acids and flavonoids of fig fruit (Ficus carica L.) in the northern Mediterranean region," Food Chemistry, vol. 106, no. 1, pp. 153–157, 2008.
16. S. Shweta, B. K. Mehta, M. Darshna, N. Hemant, and M. Aditya, "A review on pharmocoligical activity ofSyzygium cumini extracts usisng different solvents and their effective uses," International Research Journal of Pharmacy, vol. 3, pp. 54–58, 2012.
17. M. Zia-Ul-Haq, S. avar, M. Qayum, I. Imran, and V. de Feo, "Compositional studies: antioxidant and antidiabetic activities of Capparis decidua (Forsk.) Edgew," International Journal of Molecular Sciences, vol. 12, no. 12, pp. 8846–8861, 2011.
18. B. Joseph and D. Jini, "A medicinal potency of Capparis decidua—a harsh terrain plant," Research Journal of Phytochemistry, vol. 5, no. 1, pp. 1–13, 2011.
19. P. Singh, G. Mishra, S. Srivastava, K. K. Jha, and R. L. Khosa, "Traditional uses, phytochemistry and pharmacological properties of Capparis decidua: an Overview," Der Pharmacia Lettre, vol. 3, no. 2, pp. 71–82, 2011.

20. M. U. Dahot, "Chemical evaluation of the nutritive value of flowers and fruits of Capparis decidua," Journal of the Chemical Society of Pakistan, vol. 15, no. 1, pp. 78–81, 1993.
21. S. Rathee, P. Rathee, D. Rathee, and V. Kumar, "Phytochemical and pharmacological potential of kair (Capparis Decidua)," International Journal of Phytomedicine, vol. 2, no. 1, pp. 10–17, 2010.
22. L.-M. Li, X. Liao, S.-L. Peng, and L.-S. Ding, "Chemical constituents from the seeds of Ziziphus jujuba var.spinosa (Bunge) Hu," Journal of Integrative Plant Biology, vol. 47, no. 4, pp. 494–498, 2005
23. A. Sofowara, Medicinal Plants and Traditional Medicine in Africa, Spectrum Books, Ibadan, Nigeria, 1993.
24. G. E. Trease and W. C. Evans, Pharmacognosy, Braille Tirideland Macmillan Publishers, 2nd edition, 1989.
25. J. B. Harborne, Phytochemical Methods, Chapman & Hall, London, UK, 1973.
26. R. J. Grubeši , J. Vukovi , D. Kremer, and S. Vladimir-Kneževi , "Spectrophotometric method for polyphenols analysis: prevalidation and application on Plantago L. species," Journal of Pharmaceutical and Biomedical Analysis, vol. 39, no. 3-4, pp. 837–842, 2005.
27. W. Brand-Williams, M. E. Cuvelier, and C. Berset, "Use of a free radical method to evaluate antioxidant activity," LWT—Food Science and Technology, vol. 28, no. 1, pp. 25–30, 1995
28. A. Kamkar, A. J. Javan, F. Asadi, and M. Kamalinejad, "The antioxidative effect of Iranian Mentha pulegium extracts and essential oil in sunflower oil," Food and Chemical Toxicology, vol. 48, no. 7, pp. 1796–1800, 2010.
29. J. Liu, Q. Li, J. Dong, J. Chen, and G. Gu, "Multivariate modeling of aging in bottled lager beer by principal component analysis and multiple regression methods," Journal of Agricultural and Food Chemistry, vol. 56, no. 16, pp. 7106–7112, 2008.
30. C.-C. Lee, W.-C. Chang, and S.-C. Wu, "The antimutagenic and antioxidative activity for functional formula beverage," Taiwanese Journal of Agricultural Chemistry and Food Science, vol. 47, no. 6, pp. 277–284, 2009.

31. A. W. Bauer, W. M. Kirby, J. C. Sherris, and M. Turck, "Antibiotic susceptibility testing by a standardized single disk method," The American Journal of Clinical Pathology, vol. 45, no. 4, pp. 493–496, 1996.

32. AOAC (Association of Official Analytical Chemists), Official Methods of Analysis, Edited by W. Horwitz, AOAC, Arlington, Va, USA, 15th edition, 2000.

33. A. H. Gilani, M. H. Mehmood, K. H. Janbaz, A.-U. Khan, and S. A. Saeed, "Ethnopharmacological studies on antispasmodic and antiplatelet activities of Ficus carica," Journal of Ethnopharmacology, vol. 119, no. 1, pp. 1–5, 2008

34. A. Solomon, S. Golubowicz, Z. Yablowicz et al., "Antioxidant activities and anthocyanin content of fresh fruits of common fig (Ficus carica L.)," Journal of Agricultural and Food Chemistry, vol. 54, no. 20, pp. 7717–7723, 2006.

35. P. S. Benherlal and C. Arumughan, "Chemical composition and in vitro antioxidant studies on Syzygium cumini fruit," Journal of the Science of Food and Agriculture, vol. 87, no. 14, pp. 2560–2569, 2007.

36. D. T. Gordon, D. Stoops, and V. Ratliff, "Dietary fiber and mineral nutrition," in Dietary Fiber in Health & Disease, D. Kritchevsky and C. Bonfield, Eds., Eagan Press, Eagan, Minn, USA, 1995.

37. M. A. Sanjoaquin, P. N. Appleby, E. A. Spencer, and T. J. Key, "Nutrition and lifestyle in relation to bowel movement frequency: a cross-sectional study of 20630 men and women in EPIC-Oxford," Public Health Nutrition, vol. 7, no. 1, pp. 77–83, 2004.

38. D. L. Topping and P. M. Clifton, "Short-chain fatty acids and human colonic function: roles of resistant starch and nonstarch polysaccharides," Physiological Reviews, vol. 81, no. 3, pp. 1031–1064, 2001.

39. L. U. Thompson, "Potential health benefits and problems associated with antinutrients in foods," Food Research International, vol. 26, no. 2, pp. 131–149, 1993.

40. C. P'ei and S. L. Chen, "Verbenaceae," Flora Reipublicae Popularis Sinicae, vol. 65, pp. 1–229, 1982.

41. C. L. Ridout, K. R. Price, M. S. DuPont, M. L. Parker, and G. R. Fenwick, "Quinoa saponins-analysis and preliminary

investigations into the effect of reduction by processing," Journal of the Science of Food and Agriculture, vol. 54, pp. 165–176, 1991.

42. S. Borhade, "Antibacterial activity, phytochemical analysis of water extract of Syzygium cumini and analytical study by HPLC," Asian Journal of Experimental Biological Sciences, vol. 3, no. 2, pp. 320–324, 2012.
43. M. M. Cowan, "Plant products as antimicrobial agents," Clinical Microbiology Reviews, vol. 12, no. 4, pp. 564–582, 1999.
44. O. A. Sodipo, J. A. Akiniyi, and J. U. Ogunbamosu, "Studies on certain characteristics of extracts of bark of pansinystalia macruceras (K schemp) pierre Exbeille," Global Journal of Pure and Applied Sciences, vol. 6, pp. 83–87, 2000.
45. N. Salah, N. J. Miller, G. Paganga, L. Tijburg, G. P. Bolwell, and C. Rice-Evans, "Polyphenolic flavanols as scavengers of aqueous phase radicals and as chain-breaking antioxidants," Archives of Biochemistry and Biophysics, vol. 322, no. 2, pp. 339–346, 1995.
46. D. E. Okwu, "Phytochemicals and vitamin content of indigenous spices of South Eastern Nigeria," Journal of Sustainable Agriculture and the Environment, vol. 6, pp. 30–34, 2004.
47. A. Scalbert, C. Manach, C. Morand, C. Rémésy, and L. Jiménez, "Dietary polyphenols and the prevention of diseases," Critical Reviews in Food Science and Nutrition, vol. 45, no. 4, pp. 287–306, 2005.
48. K. J. Ryan and C. G. Ray, Sherris Medical Microbiology: An Introduction to Infectious Disease, McGraw-Hill, New York, NY, USA, 4th edition, 2004.
49. M. A. Bisi-Johnson and C. L. Obi, "Escherichia coli and Salmonella species: molecular landscape and therapeutic considerations: a review," Advances in Medical Sciences, vol. 1, no. 1, pp. 1–16, 2012.
50. M. K. Hellerstein, "Is chromium supplementation effective in managing type II diabetes?" Nutrition Reviews, vol. 56, no. 10, pp. 302–306, 1998

Chapter 2

The Role of Nutraceuticals in Chemoprevention and Chemotherapy and Their Clinical Outcomes

Sabita N. Saldanha[1,2], and Trygve O. Tollefsbol[1,3,4,5,6]

[1]Department of Biology, University of Alabama at Birmingham, 175A Campbell Hall, 1300 University Blvd, Birmingham, AL 35294-1170, USA

[2]Department of Math and Sciences, Alabama State University, P.O. Box 271, Montgomery, AL 36101-0271, USA

[3]Clinical Nutrition Research Center, 402 Webb Nutrition Sciences Building, 1675 University Blvd, University of Alabama at Birmingham, Birmingham, AL 35294-3360, USA

[4]Comprehensive Cancer Center, University of Alabama at Birmingham, 1802 6th Avenue South, North Pavilion 2500, Birmingham, AL 35294, USA

[5]Center for Aging, University of Alabama at Birmingham, 933 South 19th Street, Room 201, Community Health Services Building, Birmingham, AL 35294-2041, USA

[6]Nutrition Birmingham Obesity Research Center, University of Alabama at Birmingham, 402 Webb Nutrition Sciences Building, 1675 University Blvd, Birmingham, AL 35294-3360, USA

ABSTRACT

The genesis of cancer is often a slow process and the risk of developing cancer increases with age. Altering a diet that includes consumption of beneficial phytochemicals can influence the balance and availability of dietary chemopreventive agents. In chemopreventive approaches, foods containing chemicals that have anticancer properties can be supplemented in diets to prevent precancerous lesions from occurring. This necessitates further understanding of how phytochemicals can potently maintain healthy cells. Fortunately there is a plethora of plant-based phytochemicals although few of them are well studied in terms of their application as cancer chemopreventive and therapeutic agents. In this analysis we will examine phytochemicals that have strong chemopreventive and therapeutic properties in vitro as well as the design and modification of these bioactive compounds for preclinical and clinical applications. The increasing potential of combinational approaches using more than one bioactive dietary compound in chemoprevention or cancer therapy will also be evaluated. Many novel approaches to cancer prevention are on the horizon, several of which are showing great promise in saving lives in a cost-effective manner.

INTRODUCTION

The transformation of a normal cell into a cancerous phenotype requires stages of initiation, progression, and promotion by altering specific genes [1–3]. Although predisposition to cancer cannot be signaled out by a single factor, a group of factors place some individuals at a higher risk of acquiring the disease. Most of the high-risk cases may have a genetic background, but in some instances dietary choices can dictate the outcome of health. As determined by population and epidemiological studies, the predominant forms of cancer and cancer-related deaths are those of the lung and bronchus, breast, colorectal, and prostate [4, 5]. These cancers are also more prevalent in the western parts of the world and are much lower in Asian countries. A well-balanced diet that includes more of vegetables and fruits with less fat/meat intake is in most cases a staple of many Asian countries [4, 5]. Many hypotheses have supported that diet and environment greatly influence cellular function and health [6].

Phytochemicals are plant-based chemicals that mediate their positive health benefits directly, by affecting specific molecular targets such as genes, or indirectly as stabilized conjugates affecting metabolic pathways [7]. Many genes play significant roles in the cell cycle pathway, and some of these are altered in cancer cells [1, 2]. The aim of most studies is to understand and formulate mechanistic pathways by which these naturally derived chemicals can alter the fate of a cell. For a cancerous cell to survive, it should be able to proliferate, obtain energy, and establish angiogenic pathways, in a tumor mass. Altering genes that affect these pathways can serve as suitable tools to decrease tumor mass and also allow for tumor regression. In this paper, the key focus will be on mechanistic pathways that are regulated by nutraceuticals to bring about changes in the tumor environment and serve as alternative approaches for cancer prevention and therapy (Figure 1).

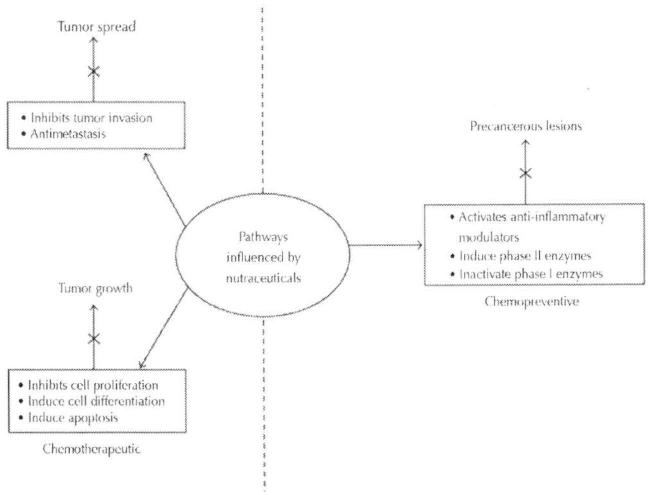

Figure 1: Cellular pathways affected by the activities of bioactive components in dietary sources. Of the natural compounds present in dietary sources, some

are more involved in regulating chemopreventive pathways and some are more effective in influencing chemotherapeutic pathways. However, a few of the bioactive molecules found to date can impart both chemopreventive and therapeutic effects, such as EGCG and genistein. Compound combinations as discussed in the paper that can affect different pathways are shown and can have profound effects on tumor growth and inhibition.

The study of phytochemicals and the classification of these compounds have been previously reviewed [8]. However, in this paper only some of the most potent and promising chemopreventive and therapeutic molecules will be analyzed, with emphasis on combination therapy of these with other nutramolecules. Most phytochemicals derived from dietary sources are classified under an umbrella of specific chemical compounds as detailed in Table 1. These molecules may not have a nutrient value but are germane to the function of a cell. Various studies have shown that these molecules can induce apoptosis, inhibit cellular proliferation, affect angiogenesis, and affect cancer metabolism in various cancers, all of which are hindrances to tumor growth (Figure 1) [7].

Table 1: Classification of nutrients as phytochemicals and their major food source availability

Phytochemical class	Bioactive compound	Source	*Molecular formula	Reference
Alkaloid	Caffeine	Cacao, tea, coffee	$C_8H_{10}N_4O_2$	[20]
	Theophylline	Cacao, tea, coffee	$C_7H_8N_4O_2$	
Monoterpenes	Limonene	Citrus oils from orange, lemon, mandarin, lime, and grapefruit	$C_{10}H_{16}$	[21]
Organosulfides	Allicin	Garlic	$C_6H_{10}OS_2$	[22–25]
	Indole-3-carbinol	Cabbage	$C_9H_{11}NO_2$	[26]
	Isothiocyanates	Broccoli	CNS	[27]
	Sulforaphane	Broccoli	$C_6H_{11}NOS_2$	[28]
Carotenoids	Beta-Carotene, lycopene	Tomatoes	$C_{40}H_{56}$	[29]

Flavonoids	Epigallocatechin-3-gallate	Green tea	C29H22O15	[30]
	Quercetin	Black tea	C15H10O7	[31]
	Curcumin	Turmeric	C21H20O6	[32]
Phenolic Acids	Capsaicin	Chilli peppers	C18H27NO3	[33]
	Ellagic acid	Black berries, raspberry	C14H6O8	[34,35]
	Gallic acid	Pomegranate, nuts	C7H6O5	[36,37]
Stilbenes	Pterostilbene	Blueberries and grapes	C16H16O3	[38]
	Resveratrol	Almonds, blueberries, grapes	C14H12O3	[39]
Isoflavones	Daidzein	Soy	C15H10O4	[9, 40]
	Genistein	Soy	C15H10O5	

*Molecular formulas obtained through the PUBCHEM COMPOUND Database.

Several of the phytochemicals listed in Table 1 have been investigated in terms of their curative properties. However, one must carefully interpret the observed results in vitro and in vivo before testing the same in a clinical setting. The reasons for this are manyfold. Tests in culture are pure, in that there is only one cell type in the culture plate and all conditions are controlled, including the bioactive compound. In vivo, however, the scenario changes as there are a host of other factors that need to be taken into account, including age, weight, diet, and metabolism of the compound. A bioactive molecule in culture may be subjected to less metabolic changes and may be presented to the cell in its native form. However, in vivo the same compound may be presented differently, perhaps as a conjugate, and its mode of action may change amongst the multitude of other molecules in the host's microenvironment. Many in vivo experiments also control for the type of diet being administered to the organism, where the concentrations or plasma availability can be adjusted. Therefore, what may work well in vitro, may have no agonistic effects or even antagonistic effects in vivo, and such discrepancies are often seen when comparing population and epidemiological studies in terms of chemical efficacy.

An effective nutraceutical is one that will have a low nontoxic dose while creating a magnitude of change in tumor dynamics. This means that at a low dose the compound should act fast on the tumor load. However, if the time taken to be effective is slow, the problems faced would be maintaining a tolerable dose and increasing bioavailability and stability. A solution to such a problem would be to use a combinatorial approach to therapy, a bioactive molecule with an effective synthetic drug or double-nutratherapy (e.g., curcumin and resveratrol). Once tumor regression sets in, dietary composition of the molecule can be adjusted.

NUTRACEUTICALS AND THEIR PREVENTIVE AND THERAPEUTIC ROLES

Genistein: a Potent Isoflavone

Many phytochemicals are currently being investigated for their promising anticarcinogenic properties. In vitroinvestigations have shown that some compounds exert their antitumor functions at much higher concentrations and that dietary consumption is insufficient to achieve such effective concentrations at the tumor site. Therefore, the mode of delivery is a very important factor that needs to be considered at clinical trials and during in vivostudies. The nontoxic properties of natural compounds are essential to the design of a formulated therapy. However, evidence along several lines of treatment has shown that some compounds are preferentially more potent in activity when administered early in life [9, 10]. For instance, soy-based prevention of breast cancer is thought to be more successful when soy products and their derivatives are consumed in early development [9].

Isoflavones are a group of phytochemicals that are predominant constituents of a soy-based diet [9, 10]. Among isoflavones, the three major constituents that have been shown to have remarkable influences in cancer prevention and therapy are genistein, diadzein, and glycitin [11]. They are collectively grouped as phytoestrogens for their weak estrogen-like activity and bind preferentially to ER-β receptors [12–15]. Evidence of antiproliferative activity of genistein in vitro stems from its ability to inhibit the tyrosine kinase enzyme that is most often upregulated in cancer cells [16, 17]. As a chemopreventive agent, genistein is thought to influence the differentiation process of mammary tissue. It is believed that early differentiation of mammary tissue into terminal buds, as seen in rats, serves as a chemopreventive strategy as it reduces the susceptibility of the epithelial cells in the ducts to carcinogens or estrogen and the ontogeny process [9]. Many aggressive cancers have altered epidermal growth factor (EGF) receptors on their cell surface allowing for a continuous downstream signaling pathway for cell division [18, 19]. This is interesting, as genistein can serve as a two-fold approach molecule for prevention and treatment. When EGF binds to its receptors, tyrosine kinase activation results in the phoshorylation of tyrosine residues of proteins involved in downstream cell signaling pathways that trigger cell division. Though studies have shown that genistein increases the EGF transcript early in development of mammary tissue, this perhaps is essential for differentiation and faster development of the breast tissue. In the long run this is a positive preventive strategy of breast lesion formation in ducts [9]. However, as seen in older rats [9], EGF mRNA decreases. Therefore, a decrease in EGF mRNA coupled with inhibition of tyrosine kinase by genistein would profoundly decrease tumor growth as cell signaling pathways are crucial to tumor maintenance.

Numerous studies have highlighted the antiproliferative role of genistein in various cancers; however, there are some studies indicating that genistein may increase cell proliferation [19, 47]. A key point to note is that nutraceuticals can be effective based on the form of genistein or its dose given at the time of the study (Tables 2and 3), especially with respect to in vitro and in vivo models. Importantly, the downstream targets of bioactive molecules under investigation need to be ascertained for each specific tissue, if overall health applications are an issue. The nutraceutical may not affect a specific common pathway for tumors of different origins. For example, in breast tissue, EGF may be highly expressed, but, in colon cells or pancreatic cells, genes that regulate cell division other than EGF may be affected [48]. Cell culture experiments using plant-based nutrients depend on the sensitivity of the cells that are being investigated. When cell lines are established, they are derived from cancerous tissues of specific organs and are, therefore, cell-type specific. This is drastically different in a clinical setting where the molecule has to mediate its activity amongst a host of various molecules and cell types. Therefore, the concentration of the phytonutrient in the supplemented diet will be crucial to its efficacy in the tumor environment. This can help explain the discrepancies seen in clinical trials of genistein for different tissues [47, 49,50]. Outcomes of some in vitro studies suggest that, like other bioactive compounds, genistein appears to have a specific cut-off concentration at which this isoflavone can exhibit anticarcinogenic activity (10 μM or even higher) [48, 51], and it is, therefore, imperative to achieve such concentrations in vivo.

Table 2: Pharmacokinetic studies evaluating the bioavailability of phytochemicals at given doses

Phyto chemical	Form	Dose	Model subject	Experimental setup	Maximum plasma concentration	Half-life (h)	Reference
Diadzein	Soy beverage	15 g Diadzein : genistein (9.27 : 10.51 mg)	Human postmenapausal women	Clinical	96.31 ng/mL	7.68	[40]
	Two soy capsules	Diadzein : enistein (7.79 : 22.57 mg)			96.02 ng/mL	6.67	
Genistein	Soy beverage	15 g Diadzein : genistein (9.27 : 10.51 mg)	Human postmenapausal women	Clinical	116.37 ng/mL	7.61	[40]
	Two soy capsules	Diadzein : genistein 7.79 : 22.57 mg)			216.84 ng/mL	7.96	
Curcumin glucoronide	Curcumoid powder form curcumin (75%), demethoxycurcumin (23%), and bisdemethoxy curcumin	10 g ()	Healthy human subjects (5 men and 7 women)	Clinical	2.04 ± 0.31	6.77 ± 0.83 for total curcumin conjugates	[41]
		12 g ()			1.40 ± 0.74		
Curcumin sulfate	Curcumoid powder form curcumin (75%), demethoxycurcumin (23%), and bisdemethoxy curcumin	10 g ()	Healthy human subjects (5 men and 7 women)	Clinical	1.06 ± 0.40	6.77 ± 0.83 for total curcumin conjugates	[41]
		12 g ()			0.87 ± 0.44		
Quercetin aglycone	Quercetin 500 plus capsule	500 mg of quercetin	Healthy human subjects (6 males and 4 female)	Clinical	15.4 ng/mL	3.47	[42]
Quercetin conjugates	Quercetin 500 plus capsule	500 mg of quercetin			336 ng/mL	Not given for plasma level, but renal clearance is 0.835	

Resveratol	Uncoated immediate-release caplets	500 mg resveratrol/caplet	Healthy human subjects	Phase I clinical test			[43, 44]
		0.5 g			72.6 (48.9)* ng/mL	2.85*	
		1.0 g			117.0 (73.1) ng/mL	8.87 (91.1)	
		2.5 g			268.0 (55.3) ng/mL	4.22 (51.6)	
		5.0 g			538.8 (72.5) ng/mL	8.52 (95.8)	
Sulforaphane	Broccoli raw	200 g	Healthy adult male subjects	Clinical	103 ± 31@, nM	3.8 ± 0.8@	[45]
	Broccoli cooked	200 g			31 ± 19@ nM	4.6 ± 0.8@	
EGCG	Beverage 200 mL	112 mg	Healthy human subjects	Clinical	Per dose (L–1) 0.51 × 10–3± 0.08 × 10–3	3.2 ± 2.1	[46]
D-Limonene oxygenated metabolite perillic acid	30–40 ounces of lemonade	447–596 mg D-limonene	Healthy human subjects	Clinical	2.08–13.98 μM	12–24	[21]
Lycopene	Lycopene with up to 250 mL water	10–120 mg	Healthy adult male subjects	Clinical	Range between 4.03 and 11.27 μg/dL (0.075–0.210 μM)	Range between 28.1 61.6 h	[29]

*coefficient of variation; @SD—standard deviation.

Table 3: Single-dose clinical studies evaluating the bioavailability of phytochemicals or their conjugated or active metabolites

Phytochemical	Route of administration	Form	**Bioavailability area under the curve (AUC)**	**AUC values**	**Reference**
Diadzein	Oral	Soy beverage	107±49.16 ng·h/mL	Adjusted to the dose	[40]
		Soy extract capsules	142.61 ± 43.94 ng·h/mL	Adjusted to the dose	[40]
Geistein	Oral	Soy beverage	121.48 ± 70.98 ng·h/mL	Adjusted to the dose	[40]
		Soy extract capsules	131.04 ± 60.79 ng·h/mL	Adjusted to the dose	[40]
Curcumin conjugates (glucoronide + sulfate)	Oral	Curcuminoid powder extract capsule form (10 g)	35.33 ± 3.78 μg·h/mL	Relative	[41]
	Oral	Curcuminoid powder extract capsule form (12 g)	26.57 ± 2.97 μg·h/mL	Relative	[41]
Quercetin aglycone	Oral	Capsule (500 mg)	62.5 ng·h/mL	Relative	[42]
Quercetin-conjugated metabolites	Oral	Capsule (500 mg)	2000 ng·h/mL	Relative	[42]

*Rersveratrol	Oral	Caplet ranging from		Relative for all	[43]
		0.5 g	223.7 ng·h/mL		
		1.0 g	544.8 (57.2) ng·h/mL		
		2.5 g	78.6 (36.2) ng·h/mL		
		5.0 g	1319 (59.1) ng·h/mL		
°Sulforaphane	Oral	200 g broccoli		Relative	[45]
		Raw	495 ± 40 nM·h		
		Cooked	286 ± 139 nM·h		
EGCG	Oral	Average 200 mL beverage	AUC	nd	[46]
D-Limonene (perillic acid a major active metabolite of d-Limonene)	Oral	40 oz of Lemonade	5.07 to 32.59 μM·h	Relative	[62]
Lycopene	Oral	Liquid form (tomato paste)	$(AUC)_{0\text{-}96}$	Relative	[29]
		10 mg	214 ± 124.8 μg·h/dL		
		30 mg	416.4 ± 183.9 μg·h/dL		
		60 mg	421.7 ± 59.3 μg.h/dL		
		90 mg	598.9 ± 396.8 μg·h/dL		
		120 mg	655 ± 298.6 μg·h/dL		

[*]AUC value measured for resveratrol was AUC infinity with the coefficient of variance denoted in the brackets against the mean value.

[δ]For the lowest dose of resveratrol AUC infinity value .

[°]AUC value measured for sulforaphane was AUC0-∞.

[κ]Based on the reference paper a list of various AUC values was given for different single doses as experimentally performed by different laboratories. Since the sample numbers were different, an average AUC value has not been given for this compound.

nd—not determined.

Isoflavones, in particular, genistein, have been extensively studied as prospective antitumor molecules in the treatment of prostate cancer [19, 52, 53]. There has been a well established line of evidence that genistein works against prostate cancer, but a majority of studies are in vitro in cultured cells [19, 52–56]. Limited clinical trials have tested the therapeutic efficacy of genistein in prostate cancer and those that have revealed inconsistencies in cell proliferation and tumor growth [57–60]. Given the inconsistencies in some of the outcomes, emphasis should be on the dose of the supplement and the form of the nutrient in the supplement at the time of administration to the patient in clinical trials. The highest achievable plasma concentration of isoflavones is 1 μM through orally administered food sources. From previous studies, this concentration is not sufficiently significant to bring about anticarcinogenic effects on the tissue. However, there is ample evidence that genistein and other isoflavones do exhibit anticancer properties and inhibit cell proliferation and tumor growth. A clinical study by Gardner et al. [61] showed that treatment of patients with dietary supplements (82 mg/day aglycone equivalents) of isoflavone yielded a higher concentration of total isoflavones in the prostatic tissues than in serum. Therefore, there is a possibility of increasing the concentration of isoflavones to anticarcinogenic levels in tissue.

An orally administered dose of isoflavones must withstand the rigors of the alimentary canal and become metabolized before they can be made available to tissues. Most isoflavones exist as conjugates rather than in their free state. This conjugation is perhaps the best way to present the molecule to the cell in tissues, and the hydrolysis of the conjugates in the tissue allows available free genistein delivery to the cells, as presented or testedin vitro. For pharmaceutical companies, it is required to formulate supplements with precise ratios of individual constituents of the compound. Unless a very pure form, a capsule or supplement may contain a mixture of genistein, diadzein, and glycetin (Tables 1, 2, 3, and 4). The percentage of each nutrient in the mixture will have a profound effect on the bioavailability of the compound after metabolism (Tables 2 and 3). To design such a product is certainly not easy and is dependent on many factors, but the two essential factors are the grade/stage of the tumor and the site or origin of the tissue. Of the two isoflavones, diadzein has been shown to have a lesser apoptotic effect on prostate cancer cells but can inhibit neoplastic transformation [61]. Therefore, it would be advantageous to use supplements containing the two bioactive nutrients as chemopreventive agents.

Table 4: Assessment of the chemotherapeutic and chemopreventive effects of nutraceuticals in combination studies

Combination of nutraceutical	Dose used	Pathways affected or mechanistic action	Organ of study	Phase of study	Model of study	Reference
Curcumin + paclitaxel	50 μM/L + 10–50 μM/L based on the gene assessed	Inactivation of NF-κB and other metastatic genes.	Breast	In vitro	Human breast cancer cells MDA-MB-435	[63]
Curcumin + paclitaxel	2% w/w 10 mg/kg	Inhibition of metastasis		In vivo	Human breast cancer xenograft model	
Curcumin + xanthorrhizol	Synergistic effect in the range from 5 to 20 μg/mL	Induction of apoptosis	Breast	In vitro	Human breast MDA-MB-231 cancer cells	[64]
Curcumin + docosahexenic acid	Ratio of DHA to CCM MCF-7 55:30 μM MCF10A 95:45 μM MDA-MB 35:35 μM SK-BR-3 60:40 μM MDA-MB 50:25 μM	Inhibition of proliferation, more synergistic in one of the 5 cell lines tested. Enhanced uptake of curcumin by the cells. Upregulated genes involved in cell cycle arrest, apoptosis, inhibition of metastasis, and cell adhesion. Downregulated genes involved in metstasis and invasion.	Breast	In vitro	Human breast cancer cells SK-BR-3, MDA-MB-231, MDA-MB-361, MCF-7, and MCF10AT	[65]

Curcumin + genistein	10 μM + 25 μM	Change in cell morphology and growth inhibition	Breast	In vitro	T47D and	[66]
	10 μM +25 μM				MCF-7	
	11 μM + 25 μM				MDA-MB-231	
Curcumin + sulphinosine	15 μM + 10 μM	Alter multidrug resistance genes.	Lung	In vitro	NCI-H460/R	[67]
		Alters the cell cycle with cells inhibited primarily in the S G2/M phase of the cycle				
Curcumin + celecoxib	10–15 μM/L + 5 μM/L	Inhibition of cell proliferation and induction of apoptosis.	Colon	In vitro	HT-29	[68]
		Possible inhibition of Cox-2 pathways or through non-Cox-2 pathways			IEC-18-K-ras (Cox-2, high levels) Caco-2 (COX-2, low levels), and SW-480 (no COX-2)	
Coltect + 5-aminosalicylic acid (ASA)	Coltect only 20 μM	Inhibition of tumor growth by induction of apoptosis.	Colon	In vitro	HT-29 cells	[69]
	150 mg/kg + 50 mg/kg	Inhibits abnormal crypt formation		In vivo	Chemical induction of tumors by 1, 2-dimethyl Hydrazine (DMH) model in rats.	
Curcumin + PEITC	25 μM + 10 μM	Additive effectives in the induction of apoptosis.	Prostate	In vitro	PC-3 C4 cell line	[27]
	3 μM + 2.5 μM	Inhibition of tumor growth through inhibition of Akt and NF- B pathways.		In vivo	NCr-immunodeficient (nu/nu) mice bearing s.c. xenografts of PC-3 human prostate cancer cells	[70]

Pure 3—curcumin + resveratrol + EGCG;	Individual compounds, Percentage composition in the diet not defined	Inhibit growth by inhibiting hedgehog signaling pathways.	Prostate	In vitro	PC-3, LnCaP and mouse cell line TRAMP-C2	[71]
Pure 4—apigenin + baicalein + genistein + quercetin; Pure 7—Pure 3 + Pure 4; Crude 7—soy + sencha leaves + turmeric + yucca roots + saw palmetto + chamomile flowers + gingko		Reduce or delay the onset of tumors.		In vivo	Transgenic adenocarcinoma of the mouse prostate (TRAMP) mice	
D-Limonene + docetaxcel	0.2 mM + 1.9 nM	Induction of apoptosis by the regulation of proteins involved in mitochondrial apoptotic pathways	Prostate	In vitro	Human prostate carcinoma DU-145 and normal prostate epithelial PZ-HPV-7 cells	[72]
Tomato powder + broccoli powder (10:10) g/100 g of diet	11 nM of lycopene per g of diet and broccoli powder, 1.6 μM of glucoraphanin, 5.9 μM of glucobrassicin, 3.9 μM of gluconasturtiin, and 2.1 μM of neogluc-obrassicin.	Reduction of tumor growth mediated by reduced cell proliferation and induction of apoptosis	Prostate	In vivo	Dunning R3327-H prostate adenocarcinoma model	[73]
Lycopene + ketosamine (fructose/amino acid Fru/His)	1 μM/L + 2 mM/L	Synergistic effect in inhibiting cell proliferation mediated processes. Antioxidant activity to prevent initiation of tumors.	Prostate	In vitro	Mat-Lylu rat cells	[74]
	20 μM/L + 5.6 mM/L	Reduce tumor growth and volume.		In vivo	Subcutaneous injections of Mat-Lylu cells in male Copenhagen rats	

Lycopene + docetaxel	1 μM + 1 nM	Synergistically enhances the antiproliferative effects of docetaxel.	Prostate	In vitro	Human PC-3, LnCaP, DU145 cells	[75]
	15 mg/kg lycopene + 10 mg/kg docetaxel	Reduced tumor volume and growth by affecting the levels of IG-FR receptor that is highly expressed in a majority of prostate tumors. Inhibited Akt signaling and suppressed surviving necessary for tumor growth		In vivo	Xenograft of DU145 cells in NCR-nu/nu (nude) mice	
Quercetin chalcone (QC) and a pH-modified citrus pectin (MCP)	1.6 mg/mL + 1.6 mg/mL	Reduction in the growth of solid primary tumors	Colon	In vivo	Balb/c mice	[76]
Quercetin + EGCG	20 μM + 0–60 μM	Inhibits the self-renewal capacity of prostate cancer stem cells (PCSCs) by synergistically inducing apoptosis decreasing cell viability in spheroids, cell migration, invasion and colony formation	Prostate	In vitro	Prostate cancer stem cells (PCSCs)	[31]
Resveratrol + estrogen (E2)	10 μM + 1 nM	Antagonistic estrogenic effects in suppression of progesterone receptor	Breast	In vitro	Human MCF-7 cells	[39]
Resveratrol + quercetin + catechin	Either all at 0.5 μM and 5 μM, or 20 μM	Synergistically inhibited cell proliferation and induced apoptosis.	Breast	In vitro	Human MDA-MB-231 cells	[77]
		Inhibited cell cycle progression with predominat cell cycle arrest in the G2 phase				
	0.5, 5, and 25 mg/kg body weight in a 100-μL volume	Reduced primary tumor growth and, therefore, inhibit tumor progression		In vivo	Breast cancer xenografts in mouse models	

Resveratrol + cyclophosphamide	50 µM + 5 mM	Inhibit cell proliferation via capase mediated cytotoxicity. Enhanced proapoptotic genes Bax, Fas and suppressed anti apoptotic gene Bcl-2	Breast	In vitro	MCF-7	[78]
Resveratrol + n-Butyrate	50 µM + 2 mM/L	Inhibited cell proliferation and induced differentiation. Attentuated p27 (Kip1) levels but enhanced p21 (Waf1/Cip1) expression.	Colon	In vitro	Caco-2	[79]
Resveratrol + 5-Fluorouracil	200 µM + IC50 800 µM	Inhibited cell proliferation and induced apoptosis by increase in capase 6 activity	Colon	In vitro	HCT116 p53 and p53	[80]
Resveratrol + genistein	250 mg/kg each in the AIN-76 diet	Suppressed prostate cancer development and mediated apoptosis by affecting the expression of steroid-receptor coactivor-3 and insulin-like growth factor-1	Prostate	In vivo	Simian Virus-40 T-antigen-(SV-40 Tag-) targeted probasin promoter rat model, a transgenic model of spontaneously developing prostate cancer.	[81]
Genistein + sulforaphane	5 µM/L + 15 µM/L	Affected DNA methyltransferase activity and reversed the gene expression of promoter hypermethylated genes of retinoic acid receptor h (RARb), RARB, p16INK4a p16 and O6-methylguanine methyltransferase enhanced growth inhibitory effects	Esophagous	In vitro	KYSE 510 cells	[82]

Sulforaphane + benzylisothio-cyanite	10 μM + 10 μM	Changed cell morphology and inhibited cell proliferation. Reduced cell viability that correlated with reduced pSTAT3 levels and an increase in PARP Cleavage	Pancreas	In vitro	PANC-1 cells	[83]
Sulforaphane + apigenin	10 μM + 10 μM	Synergistically induced phase II enzyme UDP-glucoronyl transferases (UGT1A1) transcript but to a lesser effect the protein level. Mediates this action by the induction of NF-κB	Colon	In vitro	CaCo-2	[84]
Sulforaphane + 3,3′-diindoly-lmethane (DIM)	2.5 μM + 20 μM	Has an antagonistic effect at low concentration on cell growth.	Colon	In vitro	Human colon cancer 40–16 cell line randomly derived from HCT116 clone	[85]
	Total concentration 40 μM	At cytotoxic concentrations of the compounds has synergistic effects on growth inhibition				
Sulforaphane + dibenzoymethane (DMB)	AIN-76A diet supplemented with 300 ppm SFN and 0.5% DMB	Blocked colon tumor development	Colon	In vivo	Male Apc/min + mice	[28]

Of the predominant high-risk cancers, genistein appears to have a greater affect on prostate cancers [52–54]. Genistein mediates the apoptosis of cancer cells by activating and/or inhibiting genes and/or enzymes germane to tumor maintenance (Figure 1, Table 4). Some of these important mechanisms are the inhibition of the activity of tyrosine kinase, nuclear factor kappa B (NF- B), and vitamin D 24-hydroxylase [86], activation of tumor suppressor genes, and modulation of androgen-responsive gene expression, prostate-specific antigen (PSA), and the androgen receptor (Table 4). Of the prominent isoflavonones in soy, diadzein is less effective in its action on prostate cancer, but, unlike genistein, it is metabolized to equol, an isoflavandiol which has a longer half-life than genistein [87]. The longer half- life of equol creates the possibility of using this chemical in combination with other available nutraceuticals, where the net effect may be synergistic. However, prior preclinical tests are required to investigate this possibility.

Other dietary compounds are also of great interest in this regard. In vitro, vitamin D (Vit D) has potent tumor prevention ability and can induce differentiation and apoptosis in some of the most predominant cancers [48]. The use of nutrients as a possible treatment approach is based on the fact that chemicals occurring naturally will minimize side effects when applied to a biosystem. However, the in vitro dose at which Vit D induces its antitumor properties causes hypercalcemic conditions that can preclude treatment in patients [49]. In prostate cancer, a leading cause of cancer deaths in the western parts of the world, androgen ablation therapy is the choice of treatment. However, as the cancer becomes aggressive, hormone ablation therapy fails, and progression ensues via androgen-independent pathways. Therefore, alternate therapies are very much in demand. Vitamin D is an alternate form of treatment in prostate cancer (PCA) and is shown to induce apoptosis in PCA cells in vitro. However, all PCA cell lines in vitro are not equally receptive to the vitamin D treatment or genistein [88]. Cell lines such as DU145 prostate cancer cells are especially more resistant as they express high levels of CYP24, an enzyme that catabolizes Vit D3 into less active metabolites [88]. To circumvent this problem, a recent study showed that a dual combination therapy, of DU145 to genistein and Vit D3, increased the sensitivity of the cells to Vit D3 by decreasing CYP24 expression. What is interesting to note is that the combination approach not only lowered the effective dose, but was able to abrogate cell proliferation as well. This lowered concentration of genistein at 100 nM is achieveable in vivo through dietary sources, and clinical studies would be required to determine the localization of genistein and Vit D3 in prostatic tissues.

An in vivo study for colorectal cancer has demonstrated a similar effect [89], but in this case the mice were given a single gavage of 250 μg of genistein. This mode of nutrient administration is useful for a preclinical test and probably has applications as a chemopreventive supplement. However, in terms of a clinical setting, patients are often exposed to a host of other nutrients or isoflavones in their diet, and; therefore, an in vivo model replicating such an environment with various percentages of isoflavones will allow for a better understanding of concentration and bioavailability of genistein that can mediate an apoptotic effect and reduce CYP24 expression in colonic tissues in the presence of vitamin D.

The antimetastatic properties of genistein are mediated by altering the expression of NF- B, and inhibiting the tyrosine kinase enzyme [17, 90]. Non-small-cell lung cancer (NSCLC) is a highly aggressive form of lung cancer with a poor prognosis. Therefore, alternate approaches that drastically reduce tumor growth are of utmost importance. Activation of epidermal growth factor receptor tyrosine kinase (EGFR-TK) enhances the cell signaling pathways allowing tumor growth. The use of drugs that inhibit EGFR-TK and affect NF- B, a gene whose transcribed products are essential for invasion and metastasis, can induce a more aggressive approach of reducing tumor size and the spread of the disease. A clinical therapy should be aimed at reducing tumor growth and spread by inhibiting mechanisms that contribute to the activation of metastasis. In NSCLC, genistein remarkably enhances the effects of EGFR-TK inhibitors, such as erlotinib and gefitinib, when used in combination with each of them, respectively. This effect was seen to be mediated by a marked reduction in NF- B and others, such as EGFR, pAkt, COX-2, and PGE(2), essential for regulating genes that control division, proliferation and metastasis [90]. A few studies have shown how a combined approach can lower the effective dose concentration even of chemotherapeutic drugs, minimizing potential side effects. A study conducted on breast and pancreatic cells showed that, when the cells were primed with genistein, lower concentrations of the chemotherapeutic drugs were needed to significantly bring about growth inhibition and apoptosis than with the drugs alone. In addition, NF-κB was transcriptionally inhibited in the combined treatment [90].

From a number of investigations, a common thread of evidence seems to emerge that considerable variation in the efficacy of bionutrients in cancer treatment exists and differs even among the same cell lines tested. The reasons for this are manyfold (Table 5). Cell lines derived from the same tissue hypothetically should be sensitive to the same dose or chemical class of the phytonutrients, but such is not always the case. Alternate medicinal approaches have an important task to identify crucial factors that change the sensitivity of the chemical and determine chemical modifications that would be necessary to modulate more synchronized results across several cell lines expressing similar genotypic and phenotypic signatures.

Table 5: Factors conducive to the anticarcinogenic efficacy of nutraceuticals

Factors	Possible effects on the bioactive components in the dietary supplement
Bioavailability	Metabolism
	Time taken to achieve maximum plasma concentration
	Maximum plasma concentration, half-life
Method of ingestion	Oral
	Intraperitoneal
	Subcutaneous
Form of ingestion	Powder/capsule
	Liquid
	Cooked (solid)
	Raw (solid)
Formulation	Ratio of pure to the compound conjugates
Stability	Preference for an acidic or basic environment (pH)
Mechanism of action	Direct via receptors on the cell surface or into the nuclear region via channels
	Indirect conjugated metabolites affecting parts of metabolic pathways

Epigallocatechin-3-Gallate (EGCG): a Potent Flavanol

Of the major food-derived phytochemical constituents that are extensively studied for their chemopreventive and chemotherapeutic use, EGCG and genistein are by far the most investigated. EGCG has been shown to have numerous anticancer properties which include antiangiogenic activity by affecting the transcriptional expression of vascular endothelial growth factor (VEGF) [91], inhibiting tumor initiation and promotion by inhibiting signal transduction pathways via [phosphatidyloinositol 3-kinase-Akt kinase- NF- B] [92–94], inhibiting EGFR [95], inhibiting Her-2 receptor phophorylation in breast carcinoma cells that constitutively expresses Her-2/neu receptor

[95], inducing apoptosis in estrogen receptor-(ER-) independent breast cancer cells [96], causing antimetastatic activity [97], inhibiting proteasome formation [98], inhibiting glucose-regulated protein (GRP78) activity [99]; inhibiting insulin-like growth factor-I receptor (IGF-IR) [100], and preventing invasion of tumors by inducing HMG-box transcription factor 1 (HBP1) transcriptional repressor, an inhibitor of the Wnt signaling pathway crucial for tumor-invasive property [101].

The serum level concentrations of EGCG are important to ensure that an effective response is seen without adverse or even tumor-promoting functions. Studies have shown that high doses of catechins that include a higher concentration of two prominent compounds, epicatechin gallate (ECG) and EGCG, induce hypoxia-inducible factor 1 which is responsible for activating genes related to hypoxia conditions. This allows tumor cell proliferation through alternate survival pathway mechanisms [102]. Most breast cancers are ER dependent; however, for breast cancers and others that are ER independent, EGCG inhibits the growth of tumor cells through the process of apoptosis [96, 103]. As seen in MDA-MB-468 ER-negative cells, cellular apoptosis is mediated by inducing p53 and Bax proteins that enhance apoptotic functions in cells [96]. Such observations have been corroborated by in vivo studies using animal models [97].

Most studies have shown that anticancer properties of EGCG are mediated at higher doses. However, such doses may be irrelevant to clinical applications as they may be physiologically unachievable through dietary consumption. Therefore, clinical trials should be aimed at achieving desired anticancer preventive or tumor functions at much lowered doses. Such outcomes are possible with a dual-drug approach. One study [95] demonstrated that combining EGCG with the drug taxol, which is commonly used to treat breast carcinomas, lowered the effective dose of EGCG, ranging from 0.1–1.0 μg/mL which is a serum obtainable level through metabolism. This same group showed that higher doses (30–40 μg) of EGCG were required to mediate a similar effect when used alone [95].

EGCG can be exploited as a chemopreventive agent if it prevents cancerous lesions from occurring at lower dose concentrations and for prolonged periods of time. Most in vitro studies have used relatively high doses of EGCG and such doses may prove to be more tumor promotive than preventive in longer exposure time periods. In a study designed by Pianetti et al. [92], contradictory results on the effects of EGCG on Her-2/neu overexpressed receptor in NF639 breast cancer cells was observed. At short exposure times, EGCG was very effective in reducing cell proliferation, but at prolonged exposure cells became resistant to EGCG with increased levels of NF- B. This observed change in drug-induced resistance was related to the activation of mitogen-activated protein kinase. It appears that single doses or one specific chemical constituent is mostly insufficient to induce tumor suppression or regression. Such in vitro data outcomes emphasize that a dual-drug treatment approach is necessary to treat the disease. This also signals that the timing of the nutradrug that is administered is important. Perhaps EGCG should be administered early in treatment, but later other phytochemicals or drugs, in conjunction with EGCG, may need to be administered in the treatment regimen. In their dual-drug treatment of NF639 Her-2/neu breast cancer cells, Yang et al. found that treating the cells initially with EGCG lowered cell proliferation and the later introduction of the MAPK inhibitor, U0126, reduced invasive phenotype [93].

Most studies determining the anticancer drug properties of EGCG are preclinical. For better understanding of specific EGCG effects, clinical trials should be carefully designed to include parameters that influence EGCG effectiveness. EGCG has different roles in ER-dependent versus ER-independent receptors, and, therefore, the type of diet needed to emulate in vitro doses need to be clearly understood through clinical trials and careful pharmacokinetic studies of these doses in healthy individuals, ER-positive breast cancer patients, and ER-independent tumors.

In testing phytochemicals of the same or different class it is rather uncertain which markers are necessary to determine comparable dosage values for in vitro versus in vivo efficacies. Formulation of a diet is one of the major deciding factors in the functional efficacy of a chemical constituent. It defines the concentration of the dose that will be available in vivo, after metabolism, and determines the diet that needs to be given to achieve such an outcome. Even though single-dose individual or mixed phytochemical treatments are currently available to cancer patients, they are relatively new and much more research in this direction is warranted. One such therapy that is rapidly gaining importance and holds promise for future cancer treatments is combination therapies using plant-based chemical compounds known as nutraceuticals.

COMBINATORIAL THERAPY: A PROMISE OF THE FUTURE (SEE TABLE 3)

In prevention or treatment, combinatorial approaches can be of the following types: a phytonutrient and an effective drug, two or more phytonutrients, a synthetic phytonutrient and an effective drug, or a synthetic phytonutrient and a natural nutrient. Studies in the last few decades have focused attention on unraveling the protective properties and mechanistic actions of many phytochemicals. Still the pharmacokinetics of quite a few of these phytochemicals are not known, and, for a few that are known, there is much variability based on mode and form of delivery, dose, and the model organism of study (Tables 2, 3, and 4). Another interesting approach to enhancing curative and preventive properties of these nutrients is combination therapies. The therapy is based on the factual information available at hand and using the potent properties of one with that of another to enhance synergistic or additive actions (Figure 1). In this paper, groups that have worked with different phytomolecules belonging to a different or the same chemical class of compounds have been analyzed for their antitumorigenic activities, and the overall results of the experiments for each group are described in Table 4.

Curcumin and Taxol (See [63])

Primary breast cancer cells are commonly treated with the drug taxol. Sustained chemotherapeutic treatment with this drug has often resulted in drug resistance and tumor progression. Many chemotherapeutic drugs induce the expression of the metastatic gene NF-κB which encourages tumor progression. Interestingly, natural-based compounds that are pharmacologically safe have been shown to inactivate NF-κB expression. Taxol is a powerful drug in the treatment of cancer therefore, in order to prevent metastasis, a combination of Taxol with curcumin has been shown to downregulate the expression of NF-κB and induce apoptosis.

Curcumin and Xanthorrhizol (See [64])

A study conducted on an invasive breast tumor cell line, MDA-MB-231, has shown how and when compounds added to the cells determine the overall efficacy of the treatment. A sequential addition of curcumin and xanthorrhizol (a rhizomal sesquiterpenoid of Curcuma xanthorrhiza) in culture resulted in additive and antagonistic effects depending on which compound was added first to the culture. However, simultaneous addition of the compounds resulted in synergistic effects at lower concentrations and agonistic effects at higher concentrations. Such experiments provide evidence that the efficacy of a drug is dependent on dose, time, and how it is presented to the cells. Therefore, results obtained might be contradictory if doses used are simply antagonist or additive. For a successful combination therapy or prevention, synergistic doses are more relevant to mediate downstream effects, as lower concentrations of the test biomolecules will be required.

Curcumin and Docosahexaenoic Acid (DHA) (See [65])

DHA is a dietary compound present in fish oil that has been shown to have potent chemopreventive affects against cancer. Chemotherapeutic effects of compounds are often analyzed using in vitro models. However, what is most often observed is that all cells from the same tissue sample do not react the same way to the test compound. It is essential to have a chemopreventive or therapeutic agent that can induce its effects on a wide range of cancerous cells arising from the same tissue. In this study, the authors analyzed five cell lines expressing different cell surface receptors (Table 6) which make them susceptible to chemotherapeutic compounds but in different ways and to different degrees. The combinatorial synergistic doses for each cell line were different, as shown in the Table 4. In particular, one breast cancer cell line, SK-BR-3, which is ER-negative exhibited a higher uptake of curcumin in the presence of DHA. DHA does not directly contribute to cell inhibition, but the combination of this compound with curcumin greatly enhances the uptake of curcumin by the cells. This compound, DHA, can reach a plasma concentration level of 200 μM. Although the focus of this study was entirely based on the SK-BR-3 cell line, the effects of reduced synergy on other cell lines in terms of transcriptome effects need to be investigated. Mammary tumors may contain a heterogenous population of cells exhibiting different surface receptors. Using combination therapy should be aimed at reducing the populations of all these cell types within the tumor site to truly exhibit antitumor potency with minimal side effects.

Table 6: Surface receptors expressed by breast cancer cells that alter their sensitivity to treatment

	Receptors on the surface			Cancer type	Phenotype	Reference
	ER	PR	Her2			
Breast cancer cell line						
SK-BR-3	Negative	Negative	Positive	Adenocar cinoma	Invasive	[65]
MDA-MB-231	Negative	Negative	negative	Adenocar cinoma	Invasive	[65]
MDA-MB-361	Positive	Negative	Positive	Adenocar cinoma	Metstasis	[65]
MCF-7	Positive	Positive	Negative	Adenocar cinoma	Invasive	[65]
MCF10AT	Positive	Isoform B of PR and not A	Variable	Premalig nant model for cancer develop ment	Premalignant	[65]

Curcumin and Genistein (See [66]): a Preventive Strategy

The aim to use natural compounds in diets is to render the chemopreventive properties of the compounds to the tissues. Numerous studies have shown that single dosage of compounds used alone is effective for chemoprevention. The problem faced is the inability to achieve high serum concentrations in vivo. Although combination studies are just beginning to surface as more prominent approaches in clinical treatment, studies, though limited, have shown that synergistic effects of the compounds are able to be achieved at much lower doses than when compounds are used alone. Especially in cancers that are hormonally regulated, the tissues are often exposed to external or internal hormonal stimulation, like estrogen, as in the case of breast tissue. Environmental agents that mimic estrogen-like activity can often stimulate or initiate the carcinogenic process. Curcumin, a curcuminoid, and genistein, an isoflavone, are derived from two different chemical classes, yet they have been known to inhibit a variety of tumor types in vitro and in vivo. Clinical trials of these compounds individually have been tested [19, 60, 104, 105]. The mechanistic action of the individual compounds in many different cancers has been investigated as well. However, using these compounds in combination drastically affects the development of tumors by mediating changes in shape and growth inhibition. Such changes were observed both in ER-positive and ER-negative cells, indicative of the dual use of such a combination in prevention and therapy.

Curcumin and Sulfinosine (SF) (See [67])

The ineffectiveness of certain drugs in prolonged chemotherapy stems from the resistance that some cancers develop with time. This is one of the major obstacles in cancer therapy, especially in cancers that are multidrug resistant (MDR). The problem in finding a suitable cure for non-small-cell lung cancers is the MDR phenotype it exhibits. Treating MDR cells such as non-small-cell lung carcinoma NCI-H460/R cells with a commonly employed drug, SF (obtained by the amination and subsequent oxidation of 6-thioguanosine), in cell cultures has been shown to inhibit cell growth. This observed cytotoxicity is enhanced several folds when low doses of the natural compound, curcumin, are used in combination, which are otherwise ineffective unless very high concentrations are used. These compounds mediate a synergistic effect in regulating the cell cycle phases and downregulate MDR genes, thereby, enhancing tumor regression phenotypes even in the presence of mutated p53 molecules.

Curcumin and Celecoxib (See [68])

Cyclooxygenase-2 (COX-2) expression is central to the carcinogenesis of colorectal cancers. Compounds that regulate the expression or activity of COX-2 in cells may be instrumental in mediating chemotherapeutic effects on the tissue or cells. Celecoxib is a potent inhibitor of COX-2 and is presumed to target its active site. However, prolonged exposure to celecoxib results in cardiovascular problems. It appears that monotherapy regimes are very effective in inhibiting cancer growth, proliferation, metastasis, and invasion, as seen in numerous in vitroand in vivo models. However, prolonged exposures at concentrations relatively higher than what can be achieved with combination doses may result in unwanted side effects. Testing the efficacy of celecoxib with cucumin showed that at lower doses of celecoxib it was possible to enforce synergistic inhibitory growth effects on colon cells which expressed various levels of COX-2. Like many other in vitro investigations, this study emphasizes the fact that combining powerful drugs with naturally available potent compounds can reduce the dose needed to mediate potent anticarcinogenic effects with minimal side effects. Clinically, such studies are relevant as the doses used or needed are within the physiologically dose range. With colon cancer having such a high incidence rate in the western populations, such therapies can be taken as advantage, and biomolecules having preventive potential against the formation of precancerous lesions need to be supplemented in diets of patients at high-risk.

Coltect and 5-Aminosalicylic Acid (5-ASA) (See [69])

Coltect is a novel chemotherapeutic dietary drug with a formulation of curcumin, a turmeric extract (95% curcuminoids) mixed with turmeric powder 1 : 1, green tea (60% polyphenols and 25% EGCG) in a 2 : 1 ratio, and 0.1 mg/mL of L-selenomethionine. 5-ASA is an anti-inflammatory drug, which has been shown to have a preventive role in polyp formation that is thought to occur via the inflammation process in conditions like inflammatory bowl disorder. Coltect has been effective against HT-29 human colon adenocarcinoma grade II cells in vitro, and this nutraceutical complex in combination with 5-ASA has been shown to inhibit the formation or growth of chemically induced aberrant crypt foci (ACF) in rat models. The molecular mechanism by which this inhibition is mediated is via the inhibition of COX-2 pathways in HT-29 cells, which has been supported byin vitro studies of other groups [106, 107]. However, growth inhibition can be affected via COX-2-independent pathways possibly through mechanisms that are regulated by the functional polyphenol complex in coltect. Such complex mixtures are of clinical significance as many different control mechanisms can be regulated by the presence of individual constituents of the polyphenols which are a part of the formulated mixture of coltect.

Phenylethylisothiocynate (PEITC) and Curcumin (See [27])

Most prostate cancers begin as a hormone-dependent tumor, and the hormone is primarily androgen. However, the more aggressive forms of prostate cancer are androgen-independent and hormonal therapies fail to be effective. Alternate therapies are, therefore, necessary to treat such aggressive forms. Most cancerous cells express various surface receptors that propagate cellular growth. Targeting such receptors can be an effective chemotherapeutic approach. Curcumin, obtained from Curcumin longa, has been shown to inhibit the phosphorylation of EGFR, inhibit the Akt signaling pathway, and negatively regulate NF-κB. It is an effective molecule against prostate cancer. Phenylethylisothiocyanate, a phytochemical in cruciferous vegetables, has been shown to inhibit prostate cancer cell growth in vitro and this observation has been supported by epidemiological studies showing that consumption of cruciferous vegetables has an inverse effect on prostate cancer risk. When two bioactive molecules with similar effects are used in treating hormonally independent tumors in affecting receptor mediated signaling, the effects are more pronounced than when used as individual compounds. With PEITC and curcumin, the observed effect was more additive than synergistic, but cell growth inhibition was profoundly affected by the inhibition of NF- B pathways and Akt signaling pathways. Such responses were seen at lower physiological achievable doses. These results were corroborated by in vivo studies in mice using human PC-3 prostate cancer cells [70]. Since EGCG has similar effects on prostate cancer cells, EGCG could also possibly serve as a substitute in place of curcumin for such a treatment strategy.

D-LIMONENE AND ITS COMBINATION THERAPIES (SEE TABLE 3)

Although a few studies have shown that D-limonene, an abundant monoterpene in citrus oils, exhibits antimitogenic activity, its alcohol-derivated perillyl alchohol (PA) has a greater inhibitory effect on cell migration in cancerous cells [108]. A study by Reddy et al. [108] used subtoxic doses of PA to determine this effect. Further preclinical studies are necessary to determine the effective yet nontoxic serum/tissue concentration that can be achieved from a diet rich in citrus intake, in conjunction with phytonutrients of the same class or a different class. Not much is known about the percentage composition of D-limonene and its metabolized constituents that are required to achieve an effective monterpene anticarcinogenic activity. In comparison to its oxygenated derivatives, limonene has the least cytotoxic effect on both noncancerous and cancerous breast cell lines and, therefore, can be applicable in chemoprevention [109].

D-limonene appears to be more effective against chemically induced colonic crypt foci [110]. These foci are preneoplastic lesions and are biomarkers for the progression into colon cancer. In colonic crypts that are chemically induced, limonene asserts its effect by inhibiting the activity of ornithine decarboxylase, an enzyme essential for the polyamine biosynthesis pathway. This pathway regulates the cell cycle, and D-limonene-dependent inhibition of ornithine decarboxylase (ODC) encourages an antiproliferative activity. If aberrant crypt foci are the initial markers for colon carcinogenesis, and D-Limonene and its derivatives assert their roles against initiation and promotion phases of cancer, then a diet rich in citrus foods can prevent crypt formation. Therefore, D-limonene appears to have potential as a chemopreventive agent in colon carcinogenesis. However,in vivo studies often do not correlate with results in vitro for many of the reasons discussed earlier. Once the intake of a compound is deemed safe for human consumption, it is imperative to analyze and study the mechanistic and metabolic functions in human subjects to determine the efficacy of the nutrient in question. As in the case of understanding limonene protection against colonic carcinogenesis, the studies were performed on rats and for shorter exposure time to the compound or its derivatives. Therefore, further in vivo models are required to determine the toxicity of the treatment for longer periods of time, as D-limonene is nontoxic but its alcohol derivatives could be toxic.

D-Limonene and Docetaxcel (See [72])

Many combination studies are underway to determine an effective approach in treating advanced and aggressive prostate cancers. Docetaxel, a synthetic derivative of taxol, is primarily used to inhibit the microtubular structures in cancerous cells that support cell division. In addition to its role as a microtubule disruptive molecule, it has a host of inhibitory actions on genes which regulate cell proliferation, mitotic spindle formation, transcription factors, and oncogenesis. It also upregulates genes involved in apoptosis and cell cycle progression in prostate cancer. D-Limonene, discussed earlier has been shown to have anti-prostate carcinogenic effects at low dose concentrations. Logically; therefore, combining the two compounds may have a plethora of positive antitumor functionalities. In a study by Rabi and Bishayee [72], the combined treatment enhanced the sensitivity of DU145 prostate cancer cells that are known to be apoptotic resistant. This enhanced sensitivity was thought to be mediated by reactive oxygen species (ROS) generation and activation of caspase 3 and 9. Such a positive in vitro outcome warrants further investigations in vivo, in models that mimic the progression of the disease, before it can be used in dietary supplements for therapy.

Lycopene and Fru/His (See [74, 111–114])

Serum lycopene (a carotenoid) levels have been shown to have an inverse correlation with prostate cancer risk. A diet-based population study showed that, of all the carotenoids assessed, high serum lycopene levels showed a statistically significant lower prostate cancer risk. Further analysis of their data revealed that lower serum lycopene levels in conjunction with β-carotene supplements were effective against lowering the risk of prostate cancer, suggestive for a combinatorial therapy [111]. Certain dietary compounds can be the source of cancer formation as seen with prostate cancer. It is believed that the nonfat portion of milk and excess calcium are some main factors in prostate cancer risk [112]. Numerous in vitro studies have shown that carotenoids have a greater influence in reducing tumors of the prostate origin, and lycopene and 1,2-dihydroxyvitamin D3 are at the forefront as risk reduction factors. In addition to their role as potent inhibitors of prostate cancer growth, they are biologically safe and cheaper forms of treatment. 1,2-dihydroxyvitamin D3 and lycopene have physiologically different roles, but combined they modulate pathways to synergistically inhibit proliferation and differentiation at much lower concentrations [113] and bear additive effects on cell cycle progression.

The assessments that lycopene is a safe dietary molecule with anticancer properties is supported by a number of population epidemiological and cohort-based studies [112]. However, it is important to ensure that the statistical models used are able to adjust for many parameters for a true significant outcome. Regardless of the statistical model employed in these assessment studies, lycopene has emerged as a potent risk-reducing factor of prostate cancer and has been even supported by a study that was carried out across 28 countries. Intervention combination studies have not yet been performed. However, in vivo-based studies in mice models have shown that lycopene administered in the form of tomato powder and broccoli powder in a 10 : 10 ratio, increases its serum concentration to about 538 nM/g with about 0.4 nM/g concentrated in the prostate tissue itself. Diet-based intervention studies are required to determine the formulated diet required to improve the availability in the serum of patients and enhance the localized concentration of the molecule in the tissue. Such a diet-based treatment may serve as a suitable chemopreventive approach against prostate cancer or with patients at high-risk of the disease. Even though bioactive molecules successfully work in administering their protective functionsin vitro, it appears through in vivo studies that diet and availability crucially dictate outcomes. A critical question to be asked is what factors constitute a perfect blend of bioactive mixtures. With the current research thus far, it is hard to address what the cut-off ratios are that need to be used in a diet that contain mixtures of potent nutraceuticals to coordinate similar effects clinically. Possibly a slight change in concentration of even one of the effective biomolecules may render the mixture ineffective in its function. It is rather an important task for pharmaceutical chemists and nutritionists to determine the ratios of effective biomolecules in a mixture and determine the pharmacokinetics and dynamics of that mixture.

Fru/His, a ketosamine, is also a derived product from tomatoes obtained by the reaction of a carbohydrate with an amino acid. This particular ketosamine has been found to assert chemopreventive effects by synergistically enhancing the activity of lycopene, by neutralizing ROS species and inhibiting DNA damage. Therefore, the complex of these two molecules may have a pivotal role in prostate cancer prevention. Although a rat model was used to determine the results of the treatment and pharmacokinetics of the compound are still not known, the combination of the two seemed to preferentially localize in the prostate more than in other tissues that were tested [74].

Occasionally, a combination may fail to incite anticarcinogenic effects as was seen by Mossine et al. [114]. Their experiments were conducted on the prostate adenocarcinoma rat model that was used by other groups, and their data had contradictory results to the effective action of lycopene itself and in conjunction with other micronutrients. Their study revealed that lycopene was not able to inhibit or reduce tumor load alone or in combination and that selenium alone in the mixture was able to induce antitumorigenic effects. Such outcomes are important as they open up more questions as to why a molecule that affects a given pathway behaves differently when tested within the same experimental model. Is it always dose or concentration or does molecule preparation and delivery impart effects on the efficacy of a drug?

Lycopene and Docetaxel (See [75])

Docetaxel is a potent chemotherapeutic drug that is clinically used to treat patients with advanced metastatic prostate cancers. Although the drug extends survival, it is for a very limited time period and with a poor prognosis. Lycopene, a natural compound, has been shown to have strong cancer inhibitory properties against the prostate tissue. One study tested the possibility to use this combination of compounds to enhance survival of patients that were detected with aggressive, androgen-independent tumors. As predicted, docetaxel inhibited tumor growth in nude mice that bore tumor xenografts of human DU 145 cells. Analysis of molecular mechanisms revealed that the action of docetaxel was on regulating the insulin-like growth factor receptor (IGFR) pathway by suppressing IGF, and this effect was synergistically enhanced in the presence of lycopene. Together the molecules asserted negative downstream effects on Akt signaling pathways and suppressed survivin, products of which have been known to maintain tumor growth and enhance metastasis. Clinical trials using this combination may prove effective in treating patients that express high levels of IGFR in the prostate tumor and extend survival for a longer duration than what is possibly achieved by docetaxel alone, which is about 18–20 months.

EGCG AND QUERCETIN (SEE [31])

EGCG exhibits strong chemopreventive and therapeutic activities as it influences many pathways as shown in Figure 1. Some of the mechanistic pathways are involved in regulating the levels of Bcl2, survivin, and XIAP and activation of caspase-3/7 to induce apoptosis. EGCG is also involved in inhibiting genes that are required for transition from epithelial to mesenchymal cells and retards migration and invasion which are primarily advantageous in terms of controlling aggressive tumors. EGCG mediates such synergistic actions in conjunction with quercetin to retard the self-renewal properties of cancer stem cells (CSCs), a population that, if inhibited, can influence tumor regression. Quercetin, a polyphenol, downregulates the expression of the heat shock protein (Hsp90) known to influence apoptosis and growth inhibition of prostate tumors. Therefore, the combination of these molecules modulates their respective therapeutic effects to mediate synergistic growth retardation of CSCs. The study by Tang et al. [31] used relatively higher concentrations of EGCG (60 μM) in the presence of 20 μM quercetin. Probably concentrations of EGCG that can mediate similar synergistic levels, albeit at lower doses, need to be investigated, and the therapeutic potential across cancer stems cells of other origins need to be assessed if clinical applications are to be considered.

RESVERATROL AND ESTROGEN (SEE [39])

Selective estrogen receptor modulators that are used in the clinical treatment of breast cancers display dual agonist/antagonist effects in the tissues, especially in cancer initiation and progression. Drugs like tamoxifen emulate antagonistic effects on estrogen to contain the tumor. Agonistic-estrogen-like activity can in some instances enhance tumor progression which is not desired in most clinical treatments. Resveratrol, a polyphenolic compound abundant in grape skin and grape products including wine, is known to have chemopreventive properties as supported by numerous in vitro and in vivo studies. However, based on the experimental cell type, resveratrol induces either agonistic or antagonistic effects that can be weak or very pronounced. Resveratrol agonistic effects are totally reversed in the presence of estrogen, possibly mediated through estrogen receptor β. This reversal of effects is pertinent to prevention of breast cancer lesions in ducts that could become long-term neoplastic and cancerous. Of its many cancer protective functions, resveratrol in combination with glucan are potent immunomodulators by upregulating Cdc42 expression [115]. When natural compounds exhibit multi-chemopreventive properties, conjugation therapies are advantageous over monotherapies. Albeit not clinically tested, harnessing cancer preventive and immune modulating functions of nutraceuticals seems to be a plausible approach to targeting hormonally independent aggressive tumors.

Resveratrol, Quercetin and Catechin (See Table 4 and [77])

The protective functions of polyphenols are manyfold. Numerous studies have analyzed their protective and therapeutic functions in vitro on tumor initiation that was chemically induced or in vivo via cellular implanted tumor formation. Few studies have established the functions of combined polyphenols on established tumors, as the majority of investigations have focused on individual mechanistic effects of the compounds. Dietary serum concentrations are influenced by the individual percentage of biomolecules present in the diet. Therefore, individual protective assessments of a compound show higher dose requirements, whilst mixtures may require lower doses to achieve the same effects. Additive and synergistic effects of compounds occur if their individual functions are enhanced in the presence of other molecules, perhaps by reinforcing the serum stability and availability of the various compounds in the mixture. Such observations were seen in both in vitro and in vivotesting of a mixture of three polyphenols, resveratrol, quercetin, and catechin, albeit pharmacokinetics studies are warranted.

Resveratrol and Cyclophosphamide (See [78])

Cyclophosphamide, a neoplastic drug, has a broad spectrum of activity on a variety of cancers, including breast cancers. The shortfall of the drug is its myriad of toxic effects on other systems. Dose reduction of the compound would be a means of reducing its toxicity without compensating its anticarcinogenic activity. Resveratrol has been shown to successfully lower the effective dose of cyclophosphamide without altering its anticarcinogenic activity. Both of the compounds together synergistically enhance caspase-mediated cytotoxic activity, as demonstrated in MCF-7 cells, an aggressive breast cancer cell line (Table 4). The combination therapy resulted in the upregulation of p53, proapoptotic genes, Bax and Fas, and downregulation of antiapoptotic gene Bcl-2, suggestive of an apoptotic mechanism involved in cell death.

Resveratrol and n-Butyrate (See [79])

n-Butyrate is a short chain fatty acid produced by bacterial fermentation of fiber in the colon. The compound is a known differentiating agent and induces an epithelial phenotype in certain cultured cells. n-butyrate is a potent histone deacetylase (HDAC) inhibitor as well and one of its differentiation-inducing properties stems from its ability to inhibit HDACs. Resveratrol, discussed above, induces apoptosis through other mechanistic pathways. The combination of two bioactive molecules influencing apoptosis via different mechanistic pathways may associate to render an apoptotic phenotype in cancerous cells and inhibit tumor formation and progression. The 2 mM dose of n-butyrate used in the

Wolter and Stein study [79] is probably much higher than what can be physiologically achieved. This dose is probably suitable for treatment of colon cancers where higher molar doses of n-butyrate are possible. However, n-butyrate is highly unstable, and its serum concentrations are lower than 2 mM. Since this molecule is a differentiating agent, its clinical use in treatment of other cancers is relevant. However, such therapies require absolute lower effective doses and can probably be achieved by combining with molecules other than resveratrol or modifying the compound to specific conjugates to reach serum concentration levels.

Resveratrol and 5-Fluorouracil (5-FU) (See [80])

5-fluorouracil inhibits thymidylate synthase, prevents DNA proliferation, and induces DNA damage-related apoptosis in colon cancer cells. Phase I clinical trials using a combination of resveratrol and grape powder have shown that resveratrol at low doses inhibit Wnt, a gene that is upregulated in colon cancers. Taking advantage of therapeutic effects of nutraceuticals, combined therapy of aforementioned resveratrol with 5-FU surfaces as a principal strategy in treating colon cancers. When used in combination, the presumption is that either additive or synergistic effects of the two could mediate tumor inhibition by modulating their individual apoptotic effects. The concern in using resveratrol is that higher concentrations of the doses are required in the treatment which clinically may not be reached through dietary consumption.

Genistein and Resveratrol (See [81])

Genistein and resveratrol as individual phytochemicals are very effective in the treatment and prevention of prostate cancer progression in rodent studies. Poorly differentiated prostate cancers often fail to respond to androgen-dependent treatments, and alternate treatments are required. Androgen receptors likewise have two functional roles, one as a tumor suppressor in normal prostate tissue and the other as an oncogene in neoplastic transformation, where it is altered either by mutations or DNA damage. Genistein and resveratrol used in an in vivo rat-based study, modeled to understand the mechanistic action of combined treatments in the progression of prostate cancer, showed that they had more pronounced effects, albeit not synergistic. The statistically significant additive functions of reducing cell proliferation through mechanisms that regulate the androgen receptor levels and IGF-1, a biomarker found in patient serums with progressive and aggressive prostate cancers were achieved in combined therapies over the monotherapy regimes. Interestingly, the combination of genistein and resveratrol increased serum availability of both, but higher concentrations of resveratrol were achievable as compared to the single-dose regimen. Perhaps, absorption and stability of resveratrol were profoundly affected in a combined environment, which is clinically a clear advantage. The doses used in the study are physiologically safe and achievable in vivo by consumption of a soy-based diet high in the percentage of genistein. However, resveratrol is found in low levels in grape-based dietary products, and, therefore; a pure supplement of the compound is necessary in case that higher doses are required.

GENISTEIN AND SULFORAPHANE (SFN) (SEE [82])

Previous studies have shown that EGCG, a major polyphenol in green tea, can inhibit tumor growth through mechanisms that alter DNA methylation activity, reversing the expression of silenced genes involved in tumor inhibition in cancer cells. Hypomethylation of the promoters that are CpG-rich is more likely to be transcribed, with an exception of few like hTERT, the regulatory gene of telomerase [82, 116–119]. Epigenetics is a mechanism that has been studied for decades, and factors that regulate epigenetics are now believed to be very important as treatment possibilities in controlling tumors. DNA methylation and histone deacetylation are well known epigenetic mechanisms that regulate many of the genes involved in cancers of various origins. Genistein combined with SFN, an histone deacetylase inhibitor (HDACI) has been successful in inducing the transcription of genes involved in regulating cell cycle by reversing the hypermethylated states of their promoters. This change was observed at low doses and was enhanced in the presence of sulforaphane more than that when genistein was used alone. However, in vivo studies of the same are warranted to determine epigenetic behavior of the dietary compounds before applications to human treatments are considered.

Benzylisothiocynate (BITC) and SFN (See [83])

BITC and sulforaphane are ITCs derived from cruciferous vegetables like broccoli. Individually both these molecules exert potent chemotherapeutic properties strongly supported by numerous studies. Oddly, even though both are isothiocyanates, they exert their therapeutic effects by controlling different pathways involved in tumorigenic inhibition. STAT3, a member of the STAT group of transcriptional factors, is required for early development and is dispensable in adult tissues. However, there appears to be a correlation between the constitutive expression of STAT3 and tumor development, indicative of its role as an oncogene. This gene appears to have important roles in cell proliferation, angiogenesis, and metastsis, a crucial requirement of tumor survival. Both BITC and sulforaphane have cancer inhibitory effects, affecting independent cell signaling pathways. However, the sequential combination of the two has been shown to regulate the STAT3 gene and others (Table 4), thereby, inducing apoptosis. How dietary molecules are presented to the cells in vitro is important to its cellular mechanistic actions. In the study by Hutzen et al. [83], sequential addition of BITC to the cells after sulforaphane treatment was performed, which enhanced the reduction of STAT3 levels; however, simultaneous additions were not performed. Simultaneous additions would be important for any combination study to determine possible synergistic, additive, or antagonistic effects between the compounds. Preclinical studies should include various combinatorial interactions of the nutraceuticals being tested to determine the best way of using combined molecules in therapy.

SFN and Apigenin (See [84])

Phase I and Phase II enzymes are extremely important to cancer prevention. Dietary foods are sometimes modified to produce carcinogens through metabolism by the action of Phase I enzymes. Subsequently, the action of Phase II enzymes rapidly metabolizes these products to more soluble forms that are eliminated as body waste.

Phase II enzymes are more concentrated in the duodenum and small intestine and less available in the colon. Increasing the availability of these enzymes in the colon can get rid of harmful carcinogens reducing the incidence of colon cancers, and, therefore, dietary supplements that induce Phase II enzymes would be promising tools for colon cancer prevention. SFN, an isothiocyanate, and apigenin, a flavanol, have independent cancer preventive functions. SFN is a strong inducer of UDP-27 glucuronyltransferase (UGT1A1). UGT1A1 is a major player in the detoxification process of carcinogens formed in the body and, therefore, is a potent Phase II enzyme. Treating nondifferentiated colon cells with a combination of SFN and apigenin was found to synergistically induce the expression of UGT1A1 suggesting a possible dietary tool for colon cancer prevention. The in vitro dose of the individual compounds used in the study was at physiological safe levels and can be easily achieved in vivo.

SFN and 3,3′-Diindolylmethane (DIM) (See [85])

The importance of investigating the roles of dose combinations on chemopreventive or therapeutic functions has been well dissected in a study by Pappa et al. [85]. Lower doses of SFN demonstrate antagonistic effects on cell proliferation and higher doses of both compounds had synergistic effects. Synergism of compounds is preferred if the outcome is tumor regression, but in clinical treatments synergistic actions should be mediated at safe lower concentrations rather than at cytotoxic levels. Presumably, the choice of compounds used, based on the genetic or cellular function required, is imperative to the success of the treatment. Possibly, SFN can synergistically inhibit the proliferation of cancer cells with compounds other than DIM at much lower doses, which has been investigated in studies using SFN with flavanols. This clearly highlights the problems of using combined therapies, especially since dosage is of critical importance for the success of clinical trials.

SFN and Dibenzoylmethane (DMB) (See [28])

When seeking for dietary molecules with potential chemoprotective and therapeutic properties, it is essential to understand how they mediate their combined action. Based on mechanistic studies, only compounds that are able to achieve synergistic or additive inhibitory or inductive actions on cellular genes, pathways, and/or phenotypes can then be chosen for treatment, even though their individual actions may be more pronounced. DMB is antimutagenic. Patients with aberrant polyp crypt (Apc) mutations are prone to spontaneously form aberrant polyps in their intestinal tissue, which later can transform to colorectal cancers. Treatment with DMB found in licorice can effectively inhibit such mutations in Apc, thereby protecting individuals from aberrant polyp formations. This molecule, therefore, has potential in terms of colon cancer prevention.

SFN has a myriad of chemopreventive functions as seen before in other studies and in various tissues. A combination of these two chemopreventive agents will have a profound impact on individuals that are at high-risk or reduce the incidence of colon cancers through dietary supplementation. The study by Shen et al. [28] showed that dietary intake of SFN and DMB negatively influenced the incidence and number of tumors formed in the Apc mice. The combined doses used were half that of the individual doses. However, the observed effects were still synergistic at these doses. Interestingly, the serum and plasma levels of SFN and DBM were lower in the combined doses than when the compounds were used individually. Regardless of the low serum availability, the combination was able to mediate synergistic tumor inhibitory effects. This has important clinical significance as it is possible to achieve greater tumor toxic effects at low plasma concentrations. Mechanisms that influence such actions at low serum availability need to be further investigated.

FUTURE DIRECTIONS

Chemopreventive agents are much sought after as an early interventional approach to prevent tumor development or to lower the incidence risk of cancers. Given that the current available methods of treatment are chemotherapy, radiation, and surgery, all of which can induce significant side effects, an urgent need for alternate or adjuvant therapies has arisen. Phytochemicals are relatively safe and abundantly available from dietary sources. Therefore, alternate medicine aims at harnessing the protective properties of these nonessential nutrients toward cancer prevention and treatment. A large database of studies supports the use of biomolecules in cancer treatment, albeit a majority of those are in vitro studies. Regardless of limited in vivo studies and clinical trials, phytochemicals show great promise in cancer treatment considering their safe use. Caution must be taken when addressing the efficacy of these molecules in clinical trials as many factors modulate their effects on cellular functions as detailed in Table 5. Combinatorial studies also show great promise, especially when lower nontoxic doses are required for prolonged periods to mediate potent chemotherapeutic functions with minimal side effects. Two of the major problems currently faced are dosage and delivery. To maintain a constant physiological serum dose availability, it is imperative that the agent is concentrated and stable in the tissue of concern. Combination technologies may be a solution to this problem. Nanotechnology is fast catching pace as the next level of technology in all spheres of science. Limited in vitro studies have shown that encapsulating dietary supplements in nanoparticles can effectively deliver the supplement and increase its stability and availability. Perhaps research needs to focus on such possibilities as avenues of using combination therapies.

ACKNOWLEDGMENTS

This work was supported in part by grants from the National Cancer Institute (RO1 CA129415), the American Institute for Cancer Research, and the Norma Livingston Foundation.

REFERENCES

1. W. C. Hahn, S. K. Dessain, M. W. Brooks et al., "Enumeration of the simian virus 40 early region elements necessary for human cell transformation," Molecular and Cellular Biology, vol. 22, no. 7, pp. 2111–2123, 2002.
2. W. C. Hahn, C. M. Counter, A. S. Lundberg, R. L. Beijersbergen, M. W. Brooks, and R. A. Weinberg, "Creation of human tumour cells with defined genetic elements," Nature, vol. 400, no. 6743, pp. 464–468, 1999
3. B. Elenbaas, L. Spirio, F. Koerner et al., "Human breast cancer cells generated by oncogenic transformation of primary mammary

epithelial cells," Genes and Development, vol. 15, no. 1, pp. 50–65, 2001

4. A. Jemal, M. M. Center, C. DeSantis, and E. M. Ward, "Global patterns of cancer incidence and mortality rates and trends," Cancer Epidemiology Biomarkers and Prevention, vol. 19, no. 8, pp. 1893–1907, 2010.
5. A. Jemal, F. Bray, M. M. Center, J. Ferlay, E. Ward, and D. Forman, "Global cancer statistics," CA Cancer Journal for Clinicians, vol. 61, no. 2, pp. 69–90, 2011
6. G. Fernandes, "The influence of diet and environment," Current Opinion in Immunology, vol. 2, no. 2, pp. 275–281, 1989.
7. G. J. Kelloff, J. A. Crowell, V. E. Steele et al., "Progress in cancer chemoprevention: development of diet-derived chemopreventive agents," Journal of Nutrition, vol. 130, no. 2, supplement, pp. 467S–471S, 2000
8. J. Krzyzanowska, A. Czubacka, and W. Oleszek, "Dietary phytochemicals and human health," Advances in Experimental Medicine and Biology, vol. 698, pp. 74–98, 2010
9. C. A. Lamartiniere, M. S. Cotroneo, W. A. Fritz, J. Wang, R. Mentor-Marcel, and A. Elgavish, "Genistein chemoprevention: timing and mechanisms of action in murine mammary and prostate," Journal of Nutrition, vol. 132, no. 3, pp. 552S–558S, 2002.
10. C. A. Lamartiniere, "Protection against breast cancer with genistein: a component of soy," American Journal of Clinical Nutrition, vol. 71, no. 6, supplement, pp. 1705S–1709S, 2000.
11. A. Sayeed, S. D. Konduri, W. Liu, S. Bansal, F. Li, and G. M. Das, "Estrogen receptor α inhibits p53-mediated transcriptional repression: implications for the regulation of apoptosis," Cancer Research, vol. 67, no. 16, pp. 7746–7755, 2007
12. M. Stettner, S. Kaulfuß, P. Burfeind et al., "The relevance of estrogen receptor-β expression to the antiproliferative effects observed with histone deacetylase inhibitors and phytoestrogens in prostate cancer treatment," Molecular Cancer Therapeutics, vol. 6, no. 10, pp. 2626–2633, 2007.
13. X. Tang, X. Zhu, S. Liu, R. C. Nicholson, and X. Ni, "Phytoestrogens induce differential estrogen receptor β-mediated responses in

transfected MG-63 cells," Endocrine, vol. 34, no. 1-3, pp. 29–35, 2008.

14. A. Breithofer, K. Graumann, M. S. Scicchitano, S. K. Karathanasis, T. R. Butt, and A. Jungbauer, "Regulation of human estrogen receptor by phytoestrogens in yeast and human cells," Journal of Steroid Biochemistry and Molecular Biology, vol. 67, no. 5-6, pp. 421–429, 1998

15. S. Barnes, "Phytoestrogens and breast cancer," Bailliere's Clinical Endocrinology and Metabolism, vol. 12, no. 4, pp. 559–579, 1998.

16. T. Akimoto, T. Nonaka, and H. Ishikawa, "Genistein, a tyrosine kinase inhibitor, enhanced radiosensitivity in human esophageal cancer cell lines in vitro: possible involvement of inhibition of survival signal transduction pathways," International Journal of Radiation Oncology Biology Physics, vol. 50, no. 1, pp. 195–201, 2001.

17. S. M. Gadgeel, S. Ali, P. A. Philip, A. Wozniak, and F. H. Sarkar, "Genistein enhances the effect of epidermal growth factor receptor tyrosine kinase inhibitors and inhibits nuclear factor kappa B in nonsmall cell lung cancer cell lines," Cancer, vol. 115, no. 10, pp. 2165–2176, 2009

18. H. Song, C. W. Li, A. M. Labaff et al., "Acetylation of EGF receptor contributes to tumor cell resistance to histone deacetylase inhibitors," Biochemical and Biophysical Research Communications, vol. 404, no. 1, pp. 68–73, 2011.

19. H. Nakamura, Y. Wang, T. Kurita, H. Adomat, G. R. Cunha, and Y. Wang, "Genistein increases epidermal growth factor receptor signaling and promotes tumor progression in advanced human prostate cancer,"PLoS One, vol. 6, no. 5, Article ID e20034, 2011

20. M. Li, J. Zhou, X. Gu, Y. Wang, X. J. Huang, and C. Yan, "Quantitative capillary electrophoresis and its application in analysis of alkaloids in tea, coffee, coca cola, and theophylline tablets," Journal of Separation Science, vol. 32, no. 2, pp. 267–274, 2009

21. J. Sun, "D-limonene: safety and clinical applications," Alternative Medicine Review, vol. 12, no. 3, pp. 259–264, 2007

22. S. V. Singh, "Impact of garlic organosulfides on p21H-ras processing1,2," Journal of Nutrition, vol. 131, no. 3, pp. 1046S–1048S, 2001.
23. J. A. Milner, "Mechanisms by which garlic and allyl sulfur compounds suppress carcinogen bioactivation: garlic and carcinogenesis," Advances in Experimental Medicine and Biology, vol. 492, pp. 69–81, 2001
24. J. A. Milner, "A historical perspective on garlic and cancer," Journal of Nutrition, vol. 131, no. 3, pp. 1027S–1031S, 2001
25. J. A. Milner, S. S. McDonald, D. E. Anderson, and P. Greenwald, "Molecular targets for nutrients involved with cancer prevention," Nutrition and Cancer, vol. 41, no. 1-2, pp. 1–16, 2001
26. B. B. Aggarwal and H. Ichikawa, "Molecular targets and anticancer potential of indole-3-carbinol and its derivatives," Cell Cycle, vol. 4, no. 9, pp. 1201–1215, 2005.
27. J. H. Kim, C. Xu, Y. S. Keum, B. Reddy, A. Conney, and A. N. T. Kong, "Inhibition of EGFR signaling in human prostate cancer PC-3 cells by combination treatment with β-phenylethyl isothiocyanate and curcumin," Carcinogenesis, vol. 27, no. 3, pp. 475–482, 2006
28. G. Shen, O. K. Tin, R. Hu et al., "Chemoprevention of familial adenomatous polyposis by natural dietary compounds sulforaphane and dibenzoylmethane alone and in combination in Apc Min/+ mouse," Cancer Research, vol. 67, no. 20, pp. 9937–9944, 2007.
29. D. M. Gustin, K. A. Rodvold, J. A. Sosman et al., "Single-dose pharmacokinetic study of lycopene delivered in a well-defined food-based lycopene delivery system (tomato paste-oil mixture) in healthy adult male subjects," Cancer Epidemiology Biomarkers and Prevention, vol. 13, no. 5, pp. 850–860, 2004
30. Y. H. Kao, R. A. Hiipakka, and S. Liao, "Modulation of endocrine systems and food intake by green tea epigallocatechin gallate," Endocrinology, vol. 141, no. 3, pp. 980–987, 2000
31. S. N. Tang, C. Singh, D. Nall, D. Meeker, S. Shankar, and R. K. Srivastava, "The dietary bioflavonoid quercetin synergizes with epigallocathechin gallate (EGCG) to inhibit prostate cancer stem cell characteristics, invasion, migration and epithelial-

mesenchymal transition," Journal of Molecular Signaling, vol. 5, no. 14, 2010

32. S. K. Vareed, M. Kakarala, M. T. Ruffin et al., "Pharmacokinetics of curcumin conjugate metabolites in healthy human subjects," Cancer Epidemiology Biomarkers and Prevention, vol. 17, no. 6, pp. 1411–1417, 2008.
33. X. J. Luo, J. Peng, and Y. J. Li, "Recent advances in the study on capsaicinoids and capsinoids," European Journal of Pharmacology, vol. 650, no. 1, pp. 1–7, 2011
34. H. S. Aiyer and R. C. Gupta, "Berries and ellagic acid prevent estrogen-induced mammary tumorigenesis by modulating enzymes of estrogen metabolism," Cancer Prevention Research, vol. 3, no. 6, pp. 727–737, 2010.
35. H. S. Aiyer, C. Srinivasan, and R. C. Gupta, "Dietary berries and ellagic acid diminish estrogen-mediated mammary tumorigenesis in ACI rats," Nutrition and Cancer, vol. 60, no. 2, pp. 227–234, 2008.
36. C. Y. O. Chen and J. B. Blumberg, "Phytochemical composition of nuts," Asia Pacific Journal of Clinical Nutrition, vol. 17, no. 1, pp. 329–332, 2008.
37. K. Y. Wang, "Study on the qualitative and quantitative methods of gallic acid in pomegranate rind,"Zhongguo Zhongyao Zazhi, vol. 30, no. 15, pp. 1171–1172, 2005
38. M. H. Pan, Y. H. Chang, V. Badmaev, K. Nagabhushanam, and C. T. Ho, "Pterostilbene induces apoptosis and cell cycle arrest in human gastric carcinoma cells," Journal of Agricultural and Food Chemistry, vol. 55, no. 19, pp. 7777–7785, 2007
39. K. P. L. Bhat and J. M. Pezzuto, "Resveratrol exhibits cytostatic and antiestrogenic properties with human endometrial adenocarcinoma (Ishikawa) cells," Cancer Research, vol. 61, no. 16, pp. 6137–6144, 2001.
40. E. Anupongsanugool, S. Teekachunhatean, N. Rojanasthien, S. Pongsatha, and C. Sangdee, "Pharmacokinetics of isoflavones, daidzein and genistein, after ingestion of soy beverage compared with soy extract capsules in postmenopausal Thai women," BMC Clinical Pharmacology, vol. 5, article no. 2, 2005

41. S. K. Vareed, M. Kakarala, M. T. Ruffin et al., "Pharmacokinetics of curcumin conjugate metabolites in healthy human subjects," Cancer Epidemiology Biomarkers and Prevention, vol. 17, no. 6, pp. 1411–1417, 2008
42. Y. J. Moon, L. Wang, R. DiCenzo, and M. E. Morris, "Quercetin pharmacokinetics in humans,"Biopharmaceutics and Drug Disposition, vol. 29, no. 4, pp. 205–217, 2008
43. D. J. Boocock, G. E. S. Faust, K. R. Patel et al., "Phase I dose escalation pharmacokinetic study in healthy volunteers of resveratrol, a potential cancer chemopreventive agent," Cancer Epidemiology Biomarkers and Prevention, vol. 16, no. 6, pp. 1246–1252, 2007.
44. D. J. Boocock, K. R. Patel, G. E.S. Faust et al., "Quantitation of trans-resveratrol and detection of its metabolites in human plasma and urine by high performance liquid chromatography," Journal of Chromatography B, vol. 848, no. 2, pp. 182–187, 2007.
45. M. Vermeulen, I. W.A.A. Klöpping-Ketelaars, R. Van Den Berg, and W. H.J. Vaes, "Bioavailability and kinetics of sulforaphane in humans after consumption of cooked versus raw broccoli," Journal of Agricultural and Food Chemistry, vol. 56, no. 22, pp. 10505–10509, 2008.
46. G. Williamson, F. Dionisi, and M. Renouf, "Flavanols from green tea and phenolic acids from coffee: critical quantitative evaluation of the pharmacokinetic data in humans after consumption of single doses of beverages," Molecular Nutrition and Food Research, vol. 55, no. 6, pp. 864–873, 2011.
47. D. F. Hargreaves, C. S. Potten, C. Harding et al., "Two-week dietary soy supplementation has an estrogenic effect on normal premenopausal breast," Journal of Clinical Endocrinology and Metabolism, vol. 84, no. 11, pp. 4017–4024, 1999.
48. H. S. Cross, E. Kállay, D. Lechner, W. Gerdenitsch, H. Adlercreutz, and H. J. Armbrecht, "Phytoestrogens and vitamin D metabolism: a new concept for the prevention and therapy of colorectal, prostate, and mammary carcinomas," Journal of Nutrition, vol. 134, no. 5, pp. 1207S–1212S, 2004.
49. P. G. Brooks, "Epidemiology and risk factors in breast cancer. Can we change the odds?" Journal of Reproductive Medicine for

the Obstetrician and Gynecologist, vol. 27, no. 11, pp. 670–674, 1982

50. K. B. Bouker and L. Hilakivi-Clarke, "Genistein: does it prevent or promote breast cancer?" Environmental Health Perspectives, vol. 108, no. 8, pp. 701–708, 2000.
51. P. J. Magee, H. McGlynn, and I. R. Rowland, "Differential effects of isoflavones and lignans on invasiveness of MDA-MB-231 breast cancer cells in vitro," Cancer Letters, vol. 208, no. 1, pp. 35–41, 2004
52. S. Smith, D. Sepkovic, H. L. Bradlow, and K. J. Auborn, "3,3′-Diindolylmethane and genistein decrease the adverse effects of estrogen in LNCaP and PC-3 prostate cancer cells," Journal of Nutrition, vol. 138, no. 12, pp. 2379–2385, 2008
53. R. Kumar, V. Verma, A. Sarswat et al., "Selective estrogen receptor modulators regulate stromal proliferation in human benign prostatic hyperplasia by multiple beneficial mechanisms-action of two new agents," Investigational New Drugs, pp. 1–12, 2010
54. A. Hsu, T. M. Bray, W. G. Helferich, D. R. Doerge, and E. Ho, "Differential effects of whole soy extract and soy isoflavones on apoptosis in prostate cancer cells," Experimental Biology and Medicine, vol. 235, no. 1, pp. 90–97, 2010.
55. C. S. Craft, L. Xu, D. Romero, C. P. H. Vary, and R. C. Bergan, "Genistein induces phenotypic reversion of endoglin deficiency in human prostate cancer cells," Molecular Pharmacology, vol. 73, no. 1, pp. 235–242, 2008
56. G. Zhu, W. Yan, H. C. He et al., "Inhibition of proliferation, invasion, and migration of prostate cancer cells by downregulating elongation factor-1α expression," Molecular Medicine, vol. 15, no. 11-12, pp. 363–370, 2009.
57. R. A. Jarred, M. Keikha, C. Dowling et al., "Induction of apoptosis in low to moderate-grade human prostate carcinoma by red clover-derived dietary isoflavones," Cancer Epidemiology Biomarkers and Prevention, vol. 11, no. 12, pp. 1689–1696, 2002
58. Y. Nagata, T. Sonoda, M. Mori et al., "Dietary isoflavones may protect against prostate cancer in Japanese men," Journal of Nutrition, vol. 137, no. 8, pp. 1974–1979, 2007.

59. C. J. Rosser, "Prostate cancer—to screen, or not to screen, is that the question?" BMC Urology, vol. 8, no. 1, article 20, pp. 1–3, 2008
60. R. W. deVere White, A. Tsodikov, E. C. Stapp, S. E. Soares, H. Fujii, and R. M. Hackman, "Effects of a high dose, aglycone-rich soy extract on prostate-specific antigen and serum isoflavone concentrations in men with localized prostate cancer," Nutrition and Cancer, vol. 62, no. 8, pp. 1036–1043, 2010.
61. C. D. Gardner, B. Oelrich, J. P. Liu, D. Feldman, A. A. Franke, and J. D. Brooks, "Prostatic soy isoflavone concentrations exceed serum levels after dietary supplementation," Prostate, vol. 69, no. 7, pp. 719–726, 2009.
62. H. H. S. Chow, D. Salazar, and I. A. Hakim, "Pharmacokinetics of perillic acid in humans after a single dose administration of a citrus preparation rich in d-Limonene content," Cancer Epidemiology Biomarkers and Prevention, vol. 11, no. 11, pp. 1472–1476, 2002.
63. B. B. Aggarwal, S. Shishodia, Y. Takada et al., "Curcumin suppresses the paclitaxel-induced nuclear factor-κB pathway in breast cancer cells and inhibits lung metastasis of human breast cancer in nude mice,"Clinical Cancer Research, vol. 11, no. 20, pp. 7490–7498, 2005.
64. Y. H. Cheah, F. J. Nordin, R. Sarip et al., "Combined xanthorrhizol-curcumin exhibits synergistic growth inhibitory activity via apoptosis induction in human breast cancer cells MDA-MB-231," Cancer Cell International, vol. 9, no. 1, article 1, pp. 1–12, 2009
65. J. D. Altenburg, A. A. Bieberich, C. Terry et al., "A synergistic antiproliferation effect of curcumin and docosahexaenoic acid in SK-BR-3 breast cancer cells: unique signaling not explained by the effects of either compound alone," BMC Cancer, vol. 11, no. 149, pp. 1–16, 2011
66. S. P. Verma, B. R. Goldin, and P. S. Lin, "The inhibition of the estrogenic effects of pesticides and environmental chemicals by curcumin and isoflavonoids," Environmental Health Perspectives, vol. 106, no. 12, pp. 807–812, 1998.
67. T. Andjelkovic, M. Pesic, J. Bankovic, N. Tanic, I. D. Markovic, and S. Ruzdijic, "Synergistic effects of the purine analog sulfinosine and curcumin on the multidrug resistant human non-small cell

lung carcinoma cell line (NCI-H460/R)," Cancer Biology and Therapy, vol. 7, no. 7, pp. 1024–1032, 2008

68. S. Lev-Ari, L. Strier, D. Kazanov et al., "Celecoxib and curcumin synergistically inhibit the growth of colorectal cancer cells," Clinical Cancer Research, vol. 11, no. 18, pp. 6738–6744, 2005.
69. I. Aroch, S. Kraus, I. Naumov et al., "Chemopreventive effects of Coltect, a novel dietary supplement, alone and in combination with 5-aminosalicylic acid in 1,2-dimethylhydrazine-induced colon cancer in rats,"Therapeutic Advances in Gastroenterology, vol. 3, no. 5, pp. 281–289, 2010
70. T. O. Khor, Y. S. Keum, W. Lin et al., "Combined inhibitory effects of curcumin and phenethyl isothiocyanate on the growth of human PC-3 prostate xenografts in immunodeficient mice," Cancer Research, vol. 66, no. 2, pp. 613–621, 2006.
71. A. Ślusarz, N. S. Shenouda, M. S. Sakla et al., "Common botanical compounds inhibit the hedgehog signaling pathway in prostate cancer," Cancer Research, vol. 70, no. 8, pp. 3382–3390, 2010
72. T. Rabi and A. Bishayee, "d-Limonene sensitizes docetaxel-induced cytotoxicity in human prostate cancer cells: generation of reactive oxygen species and induction of apoptosis," Journal of Carcinogenesis, vol. 8, no. 9, 2009.
73. K. Canene-Adams, B. L. Lindshield, S. Wang, E. H. Jeffery, S. K. Clinton, and J. W. Erdman, "Combinations of tomato and broccoli enhance antitumor activity in dunning R3327-H prostate adenocarcinomas,"Cancer Research, vol. 67, no. 2, pp. 836–843, 2007.
74. J. K. Campbell, K. Canene-Adams, B. L. Lindshield, T. W. M. Boileau, S. K. Clinton, and J. W. Erdman, "Tomato phytochemicals and prostate cancer risk," Journal of Nutrition, vol. 134, no. 12, supplement, pp. 3486S–3492S, 2004. View at Scopus
75. L. Tang, T. Jin, X. Zeng, and J. S. Wang, "Lycopene inhibits the growth of human androgen-independent prostate cancer cells in vitro and in BALB/c nude mice," Journal of Nutrition, vol. 135, no. 2, pp. 287–290, 2005.
76. A. Hayashi, A. C. Gillen, and J. R. Lott, "Effects of daily oral administration of quercetin chalcone and modified citrus pectin on implanted colon-25 tumor growth in balb-c mice," Alternative Medicine Review, vol. 5, no. 6, pp. 546–552, 2000.

77. A. Schlachterman, F. Valle, K. M. Wall, et al., "Combined resveratrol, quercetin, and catechin treatment reduces breast tumor growth in a nude mouse model," Translational Oncology, vol. 1, no. 1, pp. 19–27, 2008.
78. N. Singh, M. Nigam, V. Ranjan, R. Sharma, A. K. Balapure, and S. K. Rath, "Caspase mediated enhanced apoptotic action of cyclophosphamideand resveratrol-treated cf-7 cells," Journal of Pharmacological Sciences, vol. 109, no. 4, pp. 473–485, 2009.
79. F. Wolter and J. Stein, "Resveratrol enhances the differentiation induced by butyrate in Caco-2 colon cancer cells," Journal of Nutrition, vol. 132, no. 7, pp. 2082–2086, 2002
80. J. Y. Chan, S. P. Meng, M. V. Clement, S. Pervaiz, and C. L. Shao, "Resveratrol displays converse dose-related effects on 5-fluorouracilevoked colon cancer cell apoptosis: the roles of caspase-6 and p53," Cancer Biology and Therapy, vol. 7, no. 8, pp. 1305–1312, 2008.
81. C. E. Harper, L. M. Cook, B. B. Patel et al., "Genistein and resveratrol, alone and in combination, suppress prostate cancer in SV-40 tag rats," Prostate, vol. 69, no. 15, pp. 1668–1682, 2009.
82. M. Z. Fang, D. Chen, Y. Sun, Z. Jin, J. K. Christman, and C. S. Yang, "Reversal of hypermethylation and reactivation of p16INK4a, RARβ, and MGMT genes by genistein and other isoflavones from soy," Clinical Cancer Research, vol. 11, no. 19 I, pp. 7033–7041, 2005
83. B. Hutzen, W. Willis, S. Jones et al., "Dietary agent, benzyl isothiocyanate inhibits signal transducer and activator of transcription 3 phosphorylation and collaborates with sulforaphane in the growth suppression of PANC-1 cancer cells," Cancer Cell International, vol. 9, no. 24, pp. 1–7, 2009.
84. V. Svehlikova, S. Wang, J. Jakubíkova, G. Williamson, R. Mithen, and Y. Bao, "Interactions between sulforaphane and apigenin in the induction of UGT1A1 and GSTA1 in CaCo-2 cells," Carcinogenesis, vol. 25, no. 9, pp. 1629–1637, 2004
85. G. Pappa, J. Strathmann, M. Löwinger, H. Bartsch, and C. Gerhäuser, "Quantitative combination effects between sulforaphane and 3,3′-diindolylmethane on proliferation of human colon cancer cells in vitro,"Carcinogenesis, vol. 28, no. 7, pp. 1471–1477, 2007

86. Y. Li, S. Upadhyay, M. Bhuiyan, and F. H. Sarkar, "Induction of apoptosis in breast cancer cells MDA-MB-231 by genistein," Oncogene, vol. 18, no. 20, pp. 3166–3172, 1999.
87. Y. C. Chang, M. G. Nair, and J. L. Nitiss, "Metabolites of daidzein and genistein and their biological activities," Journal of Natural Products, vol. 58, no. 12, pp. 1901–1905, 1995.
88. S. Swami, A. V. Krishnan, D. M. Peehl, and D. Feldman, "Genistein potentiates the growth inhibitory effects of 1,25- dihydroxyvitamin D3 in DU145 human prostate cancer cells: role of the direct inhibition of CYP24 enzyme activity," Molecular and Cellular Endocrinology, vol. 241, no. 1-2, pp. 49–61, 2005.
89. E. Kállay, H. Adlercreutz, H. Farhan et al., "Phytoestrogens regulate vitamin D metabolism in the mouse colon: relevance for colon tumor prevention and therapy," Journal of Nutrition, vol. 132, no. 11, supplement, pp. 3490S–3493S, 2002.
90. Z. Wang, Y. Zhang, S. Banerjee, Y. Li, and F. H. Sarkar, "Inhibition of nuclear factor κB activity by genistein is mediated via Notch-1 signaling pathway in pancreatic cancer cells," International Journal of Cancer, vol. 118, no. 8, pp. 1930–1936, 2006.
91. M. R. Sartippour, Z. M. Shao, D. Heber et al., "Green tea inhibits vascular endothelial growth factor (VEGF) induction in human breast cancer cells," Journal of Nutrition, vol. 132, no. 8, pp. 2307–2311, 2002.
92. S. Pianetti, S. Guo, K. T. Kavanagh, and G. E. Sonenshein, "Green tea polyphenol epigallocatechin-3 gallate inhibits Her-2/neu signaling, proliferation, and transformed phenotype of breast cancer cells," Cancer Research, vol. 62, no. 3, pp. 652–655, 2002
93. F. Yang, H. S. Oz, S. Barve, W. J. S. De Villiers, C. J. McClain, and G. W. Varilek, "The green tea polyphenol (-)-epigallocatechin-3-gallate blocks nuclear factor-κB activation by inhibiting IκB kinase activity in the intestinal epithelial cell line IEC-6," Molecular Pharmacology, vol. 60, no. 3, pp. 528–533, 2001
94. I. A. Siddiqui, V. M. Adhami, F. Afaq, N. Ahmad, and H. Mukhtar, "Modulation of phosphatidylinositol-3-kinase/protein kinase B- and mitogen-activated protein kinase-pathways by tea polyphenols in human prostate cancer cells," Journal of Cellular Biochemistry, vol. 91, no. 2, pp. 232–242, 2004.

95. M. Masuda, M. Suzui, J. T. E. Lim, and I. B. Weinstein, "Epigallocatechin-3-gallate inhibits activation of HER-2/neu and downstream signaling pathways in human head and neck and breast carcinoma cells,"Clinical Cancer Research, vol. 9, no. 9, pp. 3486–3491, 2003.
96. A. M. Roy, M. S. Baliga, and S. K. Katiyar, "Epigallocatechin-3-gallate induces apoptosis in estrogen receptor-negative human breast carcinoma cells via modulation in protein expresssion of p53 and Bax and caspase-3 activation," Molecular Cancer Therapeutics, vol. 4, no. 1, pp. 81–90, 2005
97. M. S. Baliga, S. Meleth, and S. K. Katiyar, "Growth inhibitory and antimetastatic effect of green tea polyphenols on metastasis-specific mouse mammary carcinoma 4T1 cells in vitro and in vivo systems,"Clinical Cancer Research, vol. 11, no. 5, pp. 1918–1927, 2005
98. K. R. Landis-Piwowar, C. Huo, D. Chen et al., "A novel prodrug of the green tea polyphenol (-)-epigallocatechin-3-gallate as a potential anticancer agent," Cancer Research, vol. 67, no. 9, pp. 4303–4310, 2007.
99. S. P. Ermakova, B. S. Kang, B. Y. Choi et al., "(-)-Epigallocatechin gallate overcomes resistance to etoposide-induced cell death by targeting the molecular chaperone glucose-regulated protein 78," Cancer Research, vol. 66, no. 18, pp. 9260–9269, 2006
100. M. Li, Z. He, S. Ermakova et al., "Direct inhibition of insulin-like growth factor-I receptor kinase activity by (-)-epigallocatechin-3-gallate regulates cell transformation," Cancer Epidemiology Biomarkers and Prevention, vol. 16, no. 3, pp. 598–605, 2007
101. X. Zhang, J. Kim, R. Ruthazer et al., "The HBP1 transcriptional repressor participates in RAS-induced premature senescence," Molecular and Cellular Biology, vol. 26, no. 22, pp. 8252–8266, 2006
102. Y. D. Zhou, Y. P. Kim, X. C. Li et al., "Hypoxia-Inducible factor-1 activation by (-)-epicatechin gallate: potential adverse effects of cancer chemoprevention with high-dose green tea extracts," Journal of Natural Products, vol. 67, no. 12, pp. 2063–2069, 2004.

103. E. P. Moiseeva, L. H. Fox, L. M. Howells, L. A. F. Temple, and M. M. Manson, "Indole-3-carbinol-induced death in cancer cells involves EGFR downregulation and is exacerbated in a 3D environment," Apoptosis, vol. 11, no. 5, pp. 799–812, 2006.

104. M. Evans, J. G. Elliott, P. Sharma, R. Berman, and N. Guthrie, "The effect of synthetic genistein on menopause symptom management in healthy postmenopausal women: a multi-center, randomized, placebo-controlled study," Maturitas, vol. 68, no. 2, pp. 189–196, 2010

105. H. Ide, S. Tokiwa, K. Sakamaki et al., "Combined inhibitory effects of soy isoflavones and curcumin on the production of prostate-specific antigen," Prostate, vol. 70, no. 10, pp. 1127–1133, 2010

106. F. Rodríguez-Moranta and A. Castells, "Mechanisms of colon cancer prevention with and beyond COX-2 inhibition," Current Topics in Medicinal Chemistry, vol. 5, no. 5, pp. 505–516, 2005.

107. M. V. Swamy, C. R. Herzog, and C. V. Rao, "Inhibition of COX-2 in colon cancer cell lines by celecoxib increases the nuclear localization of active p53," Cancer Research, vol. 63, no. 17, pp. 5239–5242, 2003

108. B. S. Reddy, C. X. Wang, H. Samaha et al., "Chemoprevention of colon carcinogenesis by dietary perillyl alcohol," Cancer Research, vol. 57, no. 3, pp. 420–425, 1997

109. P. L. Crowell, A. S. Ayoubi, and Y. D. Burke, "Antitumorigenic effects of limonene and perillyl alcohol against pancreatic and breast cancer," Advances in Experimental Medicine and Biology, vol. 401, pp. 131–136, 1996.

110. T. Kawamori, T. Tanaka, Y. Hirose, M. Ohnishi, and H. Mori, "Inhibitory effects of d-limonene on the development of colonic aberrant crypt foci induced by azoxymethane in F344 rats," Carcinogenesis, vol. 17, no. 2, pp. 369–372, 1996.

111. G. M. Lowe, L. A. Booth, A. J. Young, and R. F. Bilton, "Lycopene and β-carotene protect against oxidative damage in HT29 cells at low concentrations but rapidly lose this capacity at higher doses," Free Radical Research, vol. 30, no. 2, pp. 141–151, 1999.

112. W. B. Grant, "An ecologic study of dietary links to prostate cancer," Alternative Medicine Review, vol. 4, no. 3, pp. 162–169, 1999. View at Scopus

113. D. Heber and Q. Y. Lu, "Overview of mechanisms of action of lycopene," Experimental Biology and Medicine, vol. 227, no. 10, pp. 920–923, 2002.
114. V. V. Mossine, P. Chopra, and T. P. Mawhinney, "Interaction of tomato lycopene and ketosamine against rat prostate tumorigenesis," Cancer Research, vol. 68, no. 11, pp. 4384–4391, 2008
115. V. Vetvicka, T. Volny, S. Saraswat-Ohri, A. Vashishta, Z. Vancikova, and J. Vetvickova, "Glucan and resveratrol complex–possible synergistic effects on immune system," Biomedical papers of the Medical Faculty of the University Palacký, Olomouc, Czech Republic, vol. 151, no. 1, pp. 41–46, 2007.
116. K. Jabbari and G. Bernardi, "Cytosine methylation and CpG, TpG (CpA) and TpA frequencies," Gene, vol. 333, pp. 143–149, 2004.
117. J. B. Berletch, C. Liu, W. K. Love, L. G. Andrews, S. K. Katiyar, and T. O. Tollefsbol, "Epigenetic and genetic mechanisms contribute to telomerase inhibition by EGCG," Journal of Cellular Biochemistry, vol. 103, no. 2, pp. 509–519, 2008.
118. D. Iliopoulos, P. Oikonomou, I. Messinis, and A. Tsezou, "Correlation of promoter hypermethylation in hTERT, DAPK and MGMT genes with cervical oncogenesis progression," Oncology Reports, vol. 22, no. 1, pp. 199–204, 2009.
119. D. Iliopoulos, M. Satra, A. Drakaki, G. A. Poultsides, and A. Tsezou, "Epigenetic regulation of hTERT promoter in hepatocellular carcinomas," International Journal of Oncology, vol. 34, no. 2, pp. 391–399, 2009.

Chapter 3

Research Strategies in the Study of the Pro-Oxidant Nature of Polyphenol Nutraceuticals

Harvey Babich, Alyssa G. Schuck, Jeffrey H. Weisburg, and Harriet L. Zuckerbraun

Department of Biology, Stern College for Women, Yeshiva University, 245 Lexington Avenue, New York, NY 10016, USA

ABSTRACT

Polyphenols of phytochemicals are thought to exhibit chemopreventive effects against cancer. These plant-derived antioxidant polyphenols have a dual nature, also acting as pro-oxidants, generating reactive oxygen species (ROS), and causing oxidative stress. When studying the overall cytotoxicity of polyphenols, research strategies need to distinguish the cytotoxic component derived from the polyphenol per se from that derived from the generated ROS. Such strategies include (a) identifying hallmarks of oxidative damage, such as depletion of intracellular glutathione and lipid peroxidation, (b) classical manipulations, such as polyphenol exposures in the absence and presence of antioxidant enzymes (i.e., catalase and superoxide dismutase) and of antioxidants (e.g., glutathione and N-acetylcysteine) and cotreatments with glutathione depleters, and (c) more recent manipulations, such as divalent cobalt and pyruvate to scavenge ROS. Attention also must be directed to the influence of iron and copper ions and to the level of polyphenols, which mediate oxidative stress.

INTRODUCTION

Concerns to reduce the risk of chronic diseases have led to much research to discern those lifestyle choices that potentially lessen developing chronic pathologies, cancer in particular. Epidemiological evidence has shown that the dietary consumption of fruits and vegetables can mediate the risk of developing malignancies. Studies with laboratory animal models and in vitro research with cells in culture have confirmed the anticarcinogenic effects of natural phytochemicals. In particular, the polyphenol components of phytochemicals have been identified as anticarcinogens. The most studied polyphenol is (−)-epigallocatechin-3-gallate (EGCG), the major constituent in green tea. Such nonnutritive polyphenol phytochemicals are termed nutraceuticals, and their ready bioavailability makes their consumption as potential cancer chemopreventive agents a meaningful lifestyle choice [1].

Polyphenols, a heterogeneous class of phytochemicals with a wide range of pharmacological properties, are most known for their antioxidant properties and their abilities to act as scavengers of reactive oxygen species (ROS). ROS include hydrogen peroxide (H_2O_2), superoxide anion ($O_2^{\cdot-}$), and hydroxyl radical ($OH^{\cdot}$). ROS are formed as by-products of mitochondrial respiration or by certain oxidases, such as nicotine adenine dinucleotide phosphate (NADPH) oxidase. ROS are involved in many cellular events, including as second messengers in the activation of several signaling pathways leading to the activation of transcription factors, mitogenesis, gene expression, and the induction of apoptosis, or programmed cell death [2–4]. Overproduction of ROS, as indicated by a change in the redox state of the cell, may lead to oxidative damage of proteins, lipids, and DNA. To prevent oxidative stress, neutralization of excessive ROS is accomplished by antioxidant enzymes, including superoxide dismutase (SOD) to detoxify $O_2^{\cdot-}$ and catalase and glutathione peroxidase to detoxify H_2O_2. In addition, the tripeptide, glutathione (-glutamylcysteinylglycine; GSH), plays a major role in maintaining intracellular redox balance and in alleviating ROS-induced oxidative stress. Synthesized enzymatically by -glutamylcysteine synthetase and glutathione synthetase, a prime function of GSH is to scavenge ROS and thereby to prevent oxidative damage [5]. Because oxidative stress has been implicated with cancer, as well as with other chronic diseases and pathologies, including atherosclerosis, neurodegenerative diseases, and aging, much research has focused on the antioxidant properties of plant-derived polyphenols.

Interestingly, plant-derived antioxidant polyphenols have both pro-oxidative and antioxidative properties, depending on such factors as their metal-reducing potential, chelating behavior, pH, and solubility characteristics [6]. Fukumoto and Mazza [7] noted dual antioxidant and pro-oxidant activities for a variety of plant-derived polyphenols including gallic acid, protocatechuic acid, syringic acid, vanillic acid, ellagic acid, caffeic acid, coumaric acid, chlorogenic acid, ferulic acid, myricetin, quercetin, rutin, kaempferol, (+)-catechin, (−)-epicatechin, delphinidin, and malvidin. The volume of research on the antioxidant properties of polyphenols as related to their biological effects greatly overshadows the lesser number of studies on the biological consequences of the pro-oxidant nature of polyphenols.

When interpreting cellular responses to a polyphenol, attention must be focused on the effect evoked by the polyphenol per se, as distinct from the effect evoked through its generation of significant levels of ROS. For example, Vittal et al. [8] studied gene expression in Ha-ras gene-transformed human bronchial epithelial 21BES cells exposed to 25 μM EGCG in the absence and presence of catalase, 30 U/mL. Their use of DNA microarray analyses allowed for distinction between those genes whose expressions were H_2O_2 independent from those whose expressions were H_2O_2 dependent (i.e., their expressions were abolished by catalase). Gene expression, studied at varying time intervals over a 48-hour period of exposure, indicated time-dependent expression patterns. Many of the H_2O_2-dependent genes were early response or biphasic genes associated with cell cycling, whereas, the H_2O_2-independent genes were either intermediate- or late-response genes. The initial cellular response was to H_2O_2 and, thereafter, to EGCG. The presence of catalase eliminated H_2O_2 induction of apoptotic genes (TXA2R, TNFRSF6, and MADD) and slowed down the rate of apoptosis. However, apoptotic cell death still occurred but after a delay of 12 to 24 hr, indicating that EGCG per seinduced H_2O_2-independent apoptosis.

The intent of this paper is not to review the molecular biology of the various signaling and transducing pathways ignited upon exposures to polyphenols [2, 9, 10]. Rather the goal is to discuss research strategies, some classical and others novel, to demonstrate oxidative stress as the causative agent of polyphenol-induced biological effects, in particular, antiproliferative and proapoptotic effects to cancer cells. To clarify the molecular mechanism whereby a polyphenol exerts an anticarcinogenic effect, it is important to differentiate between the polyphenol per se and its ROS auto-oxidation products.

GENERATION OF PRO-OXIDANTS

The pro-oxidant characteristic of polyphenols, as noted by their abilities to generate ROS, has been shown both in cell-free systems and in in vitro studies with cells. ROS have been detected in cell culture media and in phosphate buffers amended with polyphenols. Time-dependent generation and concentration-dependent generation of H_2O_2 were noted in Dulbeccco's modified Eagle medium (DMEM) amended with green tea, red wine [11], green tea polyphenol extract, black tea polyphenol extract [12], Ginkgo biloba extract [13], pomegranate extract [14], apple extract [15], EGCG, epigallocatechin (EGC) [12, 16], epicatechin gallate (ECG) [17], catechin gallate [18], theaflavin, theaflavin-3-monogallate, theaflavin-3′-monogallate, theaflavin-3,3′-digallate (TFdiG) [19, 20], chrysin [21], gallic acid [15, 16, 22], and quercetin [15, 16]. The quantity of H_2O_2 generated was dependent upon the specific medium. EGCG, EGC, gallic acid [16], and pomegranate extract [14] generated greater levels of ROS in DMEM, as compared to in RPMI 1640 and McCoy's media. Instability of the polyphenol at alkaline pH, resulting in its auto-oxidation, accounted for the generation of ROS in cell culture media, which most commonly was quantified by the FOX assay. The basic principle of this method is the oxidation of ferrous ions (Fe^{2+}) by the pro-oxidant polyphenol to ferric ions (Fe^{3+}), which bind with xylenol orange to give a colored complex.

The cytotoxicity of a polyphenol is dependent both on the specific polyphenol per se and upon the amount of generated ROS. Gallic acid generated considerably more H_2O_2 in DMEM than did quercetin and also exerted stronger antiproliferative activity than quercetin both to human colon Caco-2 cancer cells and to normal rat liver WB-F344 epithelial cells [22]. In DMEM, pomegranate extract generated more H_2O_2 than did olive fruit extract (i.e., a 2-hour incubation of 250 µg/mL extract in DMEM yielded 113.9±7 and 51.5±2 µmoles/L H_2O_2, resp.) and exerted greater growth inhibition to oral HSC-2 carcinoma cells (Figure1) ([14]; Schuck, unpublished).

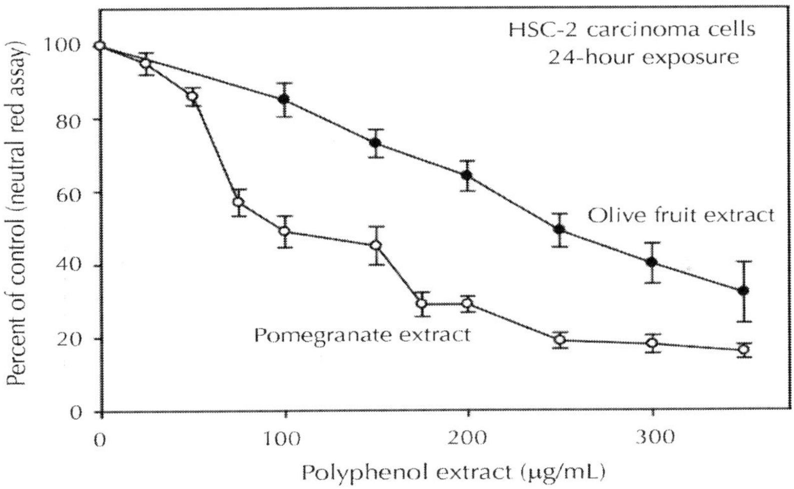

Figure 1: Comparative toxicities of pomegranate extract and olive fruit extract towards proliferation of human oral HSC-2 carcinoma cells as quantified with the neutral red assay, a cell viability assay. Data are presented as the mean percent of control ± S.E.M. Data for the pomegranate study are from Weisburg et al. [14] and for the olive fruit extract from Schuck (unpublished).

The components of the cell culture medium, including the various inorganic salts, vitamins, and amino acids, contribute to, but are not completely essential for, the generation of ROS by a polyphenol. ROS were noted in polyphenol-amended phosphate buffer although lesser amounts of ROS were detected in phosphate buffer than in cell culture medium [13, 14, 19, 22, 23]. pH is an important factor in moderating the generation of ROS by polyphenols. Generation of H_2O_2 occurred in EGCG-amended sodium phosphate buffer at pH 7.8, with lesser amounts at pH 6.8 and no detection at pH 5.8 [23]. Similar findings for the pH-mediated generation of ROS in phosphate buffer were noted with G. biloba extract (Figure 2) [13], pomegranate extract [14], and black tea theaflavins [19].

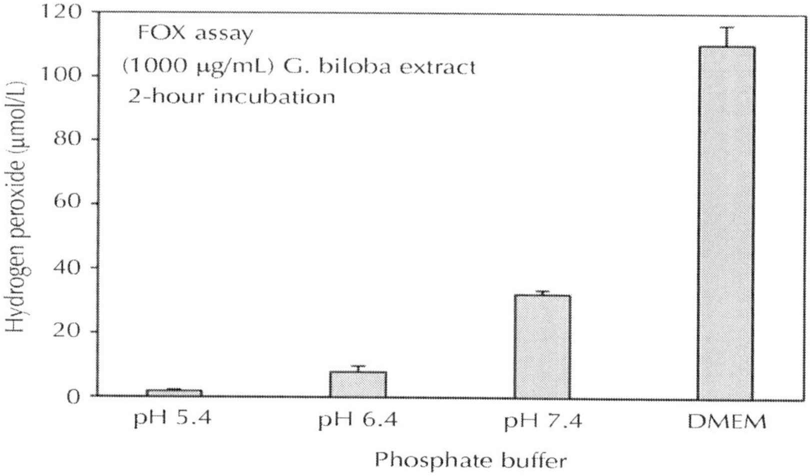

Figure 2: Comparative generation of hydrogen peroxide (H_2O_2) in phosphate buffer, maintained at different pH levels, and in cell culture medium supplemented with Gingko biloba extract. The Dulbecco's modified Eagle medium (DMEM) in this study was amended 10% Serum Plus, 2% fetal bovine serum, and antimicrobial agents and was the medium in which the cells were exposed to the test agents. H_2O_2, generated by the plant extract, was quantified with the FOX assay, after a 2-hour incubation at room temperature. Data, from Babich et al. [13], are presented as the mean percent of H_2O_2 ± S.E.M.

Another approach compared the intracellular ROS levels in unamended cells and in cells treated with polyphenols. The most common methodology utilizes the diacetate ester of 2′,7′-dichlorodihydrofluorescein (DCHF-DA), a colorless, nonfluorescent, nonpolar molecule that passively diffuses into cells. Within the cell, esterases cleave the two acetates to form DCHF, a nonpermeable, polar molecule. Oxidation of the trapped nonfluorescent DCHF by ROS, principally H_2O_2, yields the fluorescent product, 2′,7′-dichlorofluorescein. Fluorescence is quantified with a fluorometer, flow cytometer, microplate spectrophotometer, or a fluorescence microscope [24].

This methodology identified elevated levels of intracellular ROS in EGCG-treated H661 human lung cancer cells [25], leukemic UF-1 cells, freshly isolated leukemic cells from patients [26], Burkitt lymphoma HS-sultan cells, myeloma RPMI18266 cells [27], oral squamous carcinoma OSC-2 and OSC-4 cells [28], transformed human bronchial epithelial 21BES cells [29], and ovarian CAOV3 cancer cells [30], in curcumin-treated gingival fibroblasts and submandibular gland HSG carcinoma cells [31], in ovarian adenocarcinoma OVCAR-3 cervical carcinoma cells treated with ginkgetin [32], and in cactus pear extract-treated ovarian OVCA420 and SKOV3 cancer cells [33].

Whereas DCHF-DA has been used to measure intracellular ROS, particularly H_2O_2, Nakazato et al. [27] used dihydroxyethidium (DHE) to measure intracellular $O_2^{\cdot-}$ radicals. Upon uptake, cellular DHE is converted to ethidium, a fluorescent DNA intercalator, by cellular oxidants, particularly by $O_2^{\cdot-}$. Elevated levels of intracellular $O_2^{\cdot-}$ were detected in EGCG-treated lymphoma HS-sultan cells and myeloma RPMI18266 cells.

STRATEGIES TO CORRELATE CYTOTOXIC EFFECTS WITH THE PRO-OXIDANT NATURE OF A POLYPHENOL

Intracellular Glutathione

Reduced glutathione (GSH), a thiol-containing tripeptide, is a significant contributor for maintaining the intracellular redox state and, as such, is an important component of the overall cellular defensive mechanisms against ROS. An important function of this intracellular antioxidant is to scavenge ROS produced during normal aerobic cellular respiration; if left unchecked, such ROS could oxidize and, thereby, damage nucleic acids, proteins, and lipids [5].

GSH directly interacts with H_2O_2 to yield oxidized glutathione (GSSG) and H_2O. A hallmark indicator of oxidative stress is depletion of intracellular GSH (Figure 3), particularly in cancer cells which, when compared to normal cells, exhibit intrinsic oxidative stress, associated with increased metabolic activation, malfunctioning mitochondria, and oncogenic transformation [3]. Several studies have noted greater depletions of intracellular GSH in cancer, than in normal, cells upon their exposures to polyphenols, including green tea polyphenol extract [34], EGCG [12], black tea theaflavin mixture [35], theaflavin-3-gallate, theaflavin-3′-gallate [20], and TFdiG [19]. Normal cells maintain a proper intracellular redox status with their antioxidant enzymes and their sufficient supply of reduced GSH and thus are less susceptible to cytotoxic damage by pro-oxidant polyphenols [28, 30, 36].

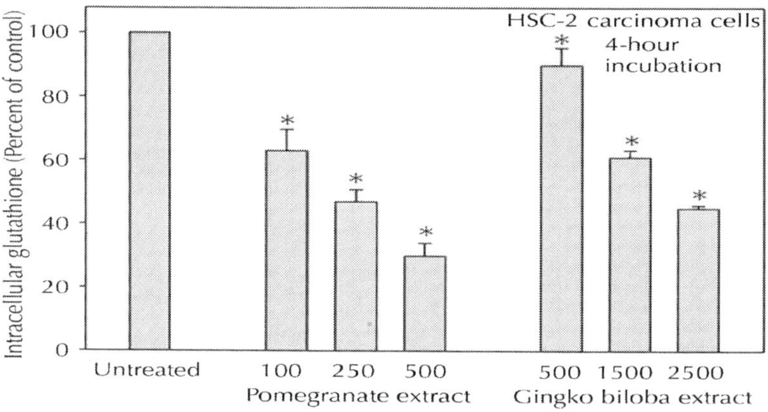

Figure 3: Decreases in intracellular reduced glutathione (GSH) following a 4-hour exposure of HSC-2 carcinoma cells in Dulbecco's modified Eagle medium amended with pomegranate extract [14] and with Ginkgo biloba extract [13]. The data are expressed as the arithmetic mean ± S.E.M.; in unexposed cells, the basal level was approximately 32 nmoles GSH/10^6 cells.

A strategy to correlate the pro-oxidant nature of a polyphenol with its cytotoxic potential is to cotreat cells with a polyphenol and a GSH depleter, as cells depleted of their intracellular GSH are hypersensitive to challenge with a pro-oxidant polyphenol. Such GSH depleters interrupt the glutathione redox cycle (GSH ↔GSSG) by inhibiting those enzymes integral to the recycling processes. Commonly used enzyme inhibitors include 1,3-bis(2-chloroethyl)-N-nitrosourea (BCNU), DL-buthionine-[S,R]-sulfoximine (BSO), and 1-chloro-2,4-dichlorobenzene (CDNB).

BCNU, an irreversible inhibitor of glutathione reductase, inhibits the recycling of intracellular GSH thereby depleting intracellular stores of the tripeptide. Cotreatment of human oral cavity cells with BCNU potentiated the antiproliferative effects of protocatechuic acid [37]. GSH is synthesized enzymatically by a two-step process, involving γ-glutamylcysteine synthetase and glutathione synthetase. BSO, a selective inhibitor of γ-glutamylcysteine synthetase, prevents the resynthesis of GSH and thereby increases cellular sensitivity to oxidative stress. Treatment of human leukemic HL-60 cells with BSO potentiated oxidative DNA damage by EGCG [38]. The antiproliferative effects of eugenol [39], EGCG [12], theaflavin-3-gallate, theaflavin-3'-gallate [20], G. biloba extract [13], and gallic acid [40] were enhanced by cotreatment with BSO. Exposure to CDNB, which serves as a substrate and covalently binds to glutathione S-transferase, irreversibly depletes intracellular GSH. Potentiation of the antiproliferative effects of eugenol [39], EGCG [12], green tea polyphenol extract [34], G. biloba extract [13], and pomegranate extract [14] was noted upon pretreatment of cells with CDNB (Figure 4).

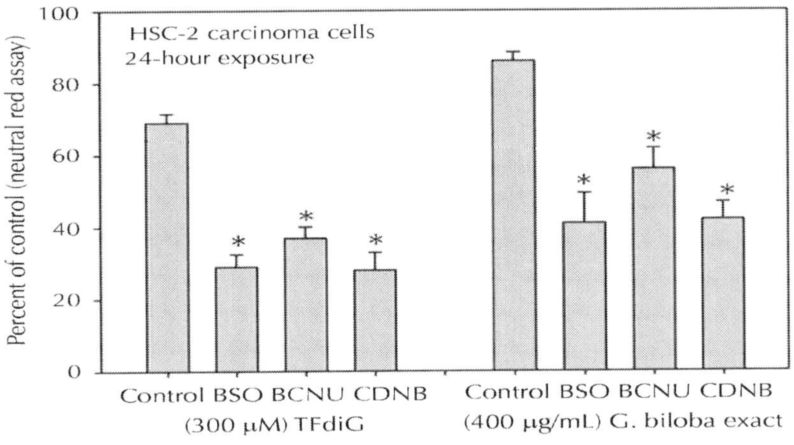

Figure 4: Comparative toxicities of theaflavin-3,3'-digallate (TFdiG) and Ginkgo bilobaextract towards human oral HSC-2 carcinoma cells, untreated (control) and treated with glutathione depleters. Cell proliferation was quantified with the neutral red assay, a cell viability assay. Data are presented as the mean percent of control ± S.E.M. Data for the TFdiG study are from Schuck et al. [19] and for G. biloba from Babich et al. [13].

Lipid Peroxidation

Another classic indicator of oxidative stress is lipid peroxidation, a process whereby free radicals extract electrons from, and thereby damage, cell membranes. Such oxidative degradation progresses via a chain reaction, initiated by ROS interacting with polyunsaturated fatty acids to generate fatty acid radicals. In the presence of O_2, a lipid peroxy fatty acid radical and thereafter a lipid peroxide radical form, with the chain reactions continuing until two free radicals interact to yield a nonradical species. Quantification of lipid peroxidation focuses on malondialdehyde (MDA), the end product of lipid peroxidation, and employs interactions between MDA and thiobarbituric acid (TBA), yielding thiobarbituric acid reactive substances (TBARSs), quantified by visible or fluorescence spectrophotometry.

Theaflavin-3-gallate and theaflavin-3′-gallate induced lipid peroxidation in human tongue CAL27 carcinoma cells [20]. Using rat hepatocyte cultures, Sahu et al. [41] demonstrated concentration-dependent lipid peroxidation upon exposure to nordihydroguaiaretic acid. Similarly, lipid peroxidation of freshly isolated lymphocytes was observed upon treatments with caffeic acid and gallic acid [42] and of erythrocytes by G. biloba extract [43]. Induction of lipid peroxidation by pro-oxidant polyphenols has been an underutilized assay.

Ros Scavenging Enzymes

Catalase enzymatically degrades H_2O_2 [$2H_2O_2 \rightarrow 2H_2O + O_2$]. By far, the most common method to demonstrate that a polyphenol-induced cytotoxic response was due to the generation of H_2O_2 is to assess that adverse cellular response in the absence and presence of added catalase. The complete abolition of a cytotoxic effect of a polyphenol through cotreatment with catalase, apparently, would indicate that the adverse effect was mediated solely through the generation of H_2O_2. For example, the growth inhibitory and apoptotic-inducing effects of EGCG towards lymphoblastic leukemic Jurkat cells were completely suppressed by addition of catalase [23]. Coexposure of lymphoma HS-sultan cells to catalase completely prevented EGCG-induced apoptosis; downregulation of the antiapoptotic-associated proteins, Bcl-2 and Mcl-1, and of procaspase-3 after EGCG treatment was prevented by catalase pretreatment [27]. Yang et al. [25, 29] noted that the EGCG-induced apoptosis and elevated intracellular ROS in human colon H661 cancer cells were completely abolished by exogenously added catalase.

Many studies, however, showed that catalase afforded only a partial protection towards pro-oxidant polyphenol toxicities. Exposure of ovarian carcinoma OVCAR-3 to ginkgetin resulted in elevated intracellular ROS, growth inhibition, and apoptosis. Catalase afforded partial protection against growth inhibition and apoptosis, as noted by reductions in DNA fragmentation and double-stranded DNA breakage [32]. The antiproliferative effects of theaflavin-3-gallate, theaflavin-3'-gallate [20], TFdiG [19], G. bilobaextract [13], and EGCG [12] to oral carcinoma cells were significantly lessened, but not completely abolished, in the presence of catalase (Figure 5).

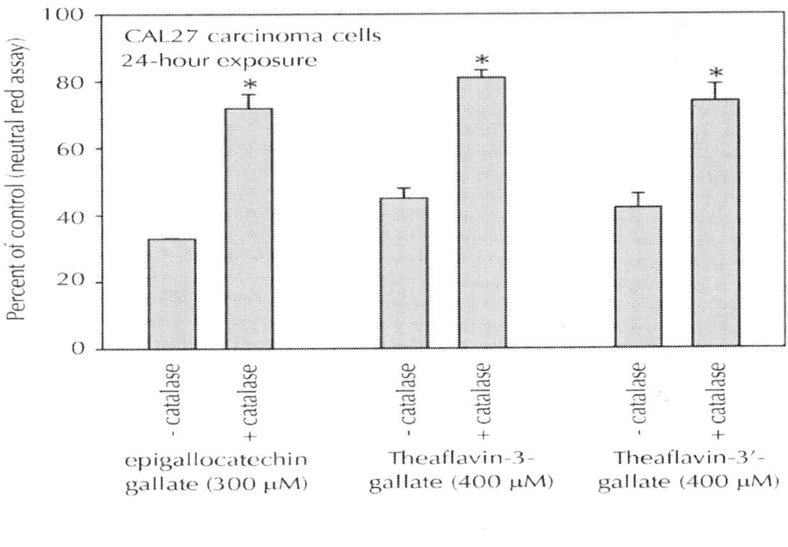

Figure 5: Comparative toxicities of epigallocatechin gallate (EGCG), theaflavin-3-digallate, and theaflavin-3'-gallate extract towards human oral HSC-2 carcinoma cells in the absence and presence of exogenously added catalase. Cell proliferation was quantified with the neutral red assay, a cell viability assay. Data are presented as the mean percent of control ± S.E.M. Data for EGCG are from Weisburg et al. [12] and for the black tea theaflavin monogallates are from Babich et al. [20].

In their studies on the response of human lung tumor H441 cells to EGCG, Yang et al. [25] noted that at similar EGCG concentrations, coexposures with exogenous catalase completely prevented apoptosis but only partially blocked the inhibition of cell proliferation. In subsequent studies with 21BES cells, Yang et al. [29] showed that exposures to TFdiG, EGCG, and EGC resulted in equivalent levels of intracellular H_2O_2 and in the induction of apoptosis. Yet, exogenously added catalase significantly prevented EGC- and EGCG-induced apoptosis but did not prevent TFdiG-induced apoptosis. Apparently, for these cells, EGCG and EGC induced apoptosis via oxidative stress, whereas TFdiG per se induced apoptosis by an ROS-independent mechanism.

EGCG-generated ROS in cell culture medium, and elevated levels of intracellular ROS were detected in human oral OSC-2 and OSC-4 EGCG-treated carcinoma cells. However, neither exogenous catalase nor N-acetylcysteine (NAC), an antioxidant, rescued the cells from the antiproliferative effects of EGCG [28]. Similarly, neither catalase nor NAC lessened the antiproliferative effect of (−)-catechin gallate to transformed oral S-G cells [18]. Apparently for these oral carcinoma cells, EGCG per se and catechin gallate per se, rather than their auto-oxidation ROS products, accounted for growth inhibition. However, Yamamoto et al. [28] observed that exogenous catalase partially rescued OCS-2 cells and more substantially rescued OSC-4 cells from EGCG-induced apoptotic cell death, as assayed by caspase-3 activation. Apparently, for the OSC-2 and OSC-4, the antiproliferative effects were due to EGCG per se, whereas the proapoptotic effects were due to EGCG-generated ROS.

Lee et al. [22] studied the effects of gallic acid on gap-junction intracellular communication (GJIC), a process essential for maintaining homeostatic balance by modulating cell growth and differentiation and whose inhibition was linked to tumor promotion. GJIC in normal rat liver epithelial WB-344 cells was inhibited by the addition of authentic H_2O_2 and of gallic acid, a strong generator of H_2O_2. For H_2O_2, inhibition of GJIC was completely abolished by the addition of catalase, but for gallic acid, catalase only partially abolished this inhibition. Apparently, inhibition of GJIC by gallic acid was attributable both to the polyphenol per se and to H_2O_2.

Cell culture medium amended with either green tea or with red wine inhibited proliferation of rat pheochromocytoma PC12 cells. Both amendments generated H_2O_2, and the addition of catalase completely abolished the antiproliferative effects of green tea, but only partially reduced that of red wine. Amendment with exogenous GSH had no significant effect on red wine toxicity. Apparently, H_2O_2 accounted for the total cytotoxic effect of green tea, but only partially for that of red wine [11]. It was suggested that the overall toxicity of red wine was a combination of H_2O_2 and of resveratrol, the main polyphenol in red wine, which has known antiproliferative effects [44].

Apparently, the mediating effects of exogenous catalase on polyphenol cytotoxicity are varied. The complete abolition of a polyphenol-induced cytotoxic effect by exogenously added catalase indicates that the generated H_2O_2 alone was the toxic agent. Partial blockage of the polyphenol-induced cytotoxic effect by catalase indicates a dual mode of toxicity, the polyphenol per se in conjunction with its auto-oxidation ROS products. No lessening of cytotoxicity in the presence of catalase presumably is indicative of the polyphenolper se alone acting as the cytotoxic agent. However, other factors may have been involved, as the methodologies used in these studies may have had unforeseen effects on the biologic responses. For example, Vittal et al. [8] showed that cellular responses to EGCG included early responding genes whose expressions were H_2O_2 dependent and latter responding genes whose expressions were dependent upon EGCG per se(i.e., were H_2O_2 independent). The early responding genes were activated soon after exposure to EGCG, as early as 15 minutes afterexposure and with increased gene expression returning to basal levels by 3 to 6 hrs. A second cascade of gene expression followed between 3 and 10 hrs and a later response clustered between 12 and 36 hrs afterEGCG treatment. Thus, the time period of cell exposure to a polyphenol would be critical in evaluating the mediating influence of catalase. In many in vitro studies, the exposure period is 24 hr or less; in such a limited exposure period, attention would be focused towards H_2O_2-dependent genes, favoring their mediation by the addition of catalase. Another factor to consider is the pyruvate content of the cell culture medium, as it is a scavenger of H_2O_2. (The mediating influence of pyruvate is discussed in a later section). Finally, the toxicity of a polyphenol may be dependent upon ROS other than H_2O_2, in particular upon $O_2^{\cdot-}$, whose activity is unaffected by catalase, but rather by SOD. As such, to rule out ROS as mediators of a polyphenol-induced cytotoxic effect, cell exposures to the polyphenol need to be done in the absence of antioxidant enzymes and in presence of catalase, of SOD, and of a combination of catalase and SOD.

As compared to catalase, fewer studies have been directed towards the effects of exogenous SOD on the toxic potency of pro-oxidant polyphenols. Superoxide radical is converted to H_2O_2, either spontaneously or enzymatically via SOD. Under physiological conditions, SOD, which requires metal cofactors, catalyzes the dismutation of the superoxide radical [$2O_2^{\cdot -} + 2H^+ \rightarrow O_2 + H_2O_2$]. Elevation of intracellular $O_2^{\cdot -}$ was observed in EGCG-treated lymphoma and myeloma cells [27]. Treatment of HL-60 leukemia cells with EGCG resulted in elevated intracellular ROS, with a concomitant increase in DNA damage, as detected in the alkaline comet assay. Coexposures either with SOD or with catalase largely prevented EGCG-mediated DNA damage and greatly reduced ROS formation; combinations of SOD and catalase completely inhibited intracellular ROS formation and DNA damage [45].

The generation of $O_2^{\cdot -}$ may be a major component of EGCG- [46] and of theaflavin-mediated toxicity, as Yen et al. [47] noted that DNA damage to EGCG-, EGC-, and theaflavin-treated Chang liver cells was associated with the generation of $O_2^{\cdot -}$. In a cell-free system, SOD lessened the EGCG-mediated generation of H_2O_2 in sodium phosphate buffer (pH 7.8) and in RPMI medium, as well as suppressing the cytotoxicity of EGCG to Jurkat cells [23]. DNA breakage, as quantified in the comet assay, in human peripheral lymphocytes treated with resveratrol was inhibited by treatment with SOD [48].

Antioxidants

N-acetylcysteine (NAC), a derivative of the amino acid, L-cysteine with an acetyl group attached to the N atom, is a precursor in the synthesis of glutathione. NAC and GSH are typically used as exogenously added antioxidants to lessen the potency of pro-oxidant polyphenols. Using oral cancer cells as the bioindicators and inhibition of cell proliferation as the cytotoxicity endpoint, the potencies of EGCG [49] and of G. bilobaextract [13] were lessened by NAC, of curcumin by NAC and GSH [50], and of green tea polyphenol extract by GSH [34]. Treatment of myeloid leukemic UF-1 cells with EGCG resulted in elevated intracellular ROS and in the induction of apoptosis, both of which were blocked upon coexposure with NAC [26]. Curcumin-induced onset of early apoptosis, as detected by the externalization of phosphatidylserine on cell surfaces of normal human gingival fibroblasts and human submandibular gland HSG carcinoma cells, and of the generation of intracellular H_2O_2, were lessened upon coexposures with exogenous NAC and GSH [31]. EGCG-induced activation of mitogen-activated protein kinases (MAPKs), release of cytochrome c, and apoptotic cell death of human colorectal HT-29 cells were significantly blocked in the presence of GSH and NAC. Interestingly, pretreatment with catalase failed to lessen MAPK activation and apoptotic cell death, indicating the involvement of ROS other than H_2O_2 [51].

Pyruvate

Pyruvate, amended to cell culture medium as sodium pyruvate, is a scavenger of H_2O_2; it nonenzymatically participates in a direct oxidative decarboxylation with H_2O_2 to yield acetate, carbon dioxide, and water [$CH_3COCOO^- + H_2O_2 \rightarrow CH_3COO^- + CO_2 + H_2O$] [52–54]. As such, pyruvate affords protection against pro-oxidant polyphenols. Exposure of human ovarian SKOV3 and CAOV3 cancer cells to EGCG resulted in growth inhibition, accompanied by increased intracellular levels of H_2O_2. Addition of pyruvate to the culture medium neutralized the cytotoxicity of EGCG [30].

The antiproliferative toxicities of TFdiG [19], G. biloba extract [13], and pomegranate extract [14] to oral carcinoma HSC-2 cells were lessened in the presence of exogenous pyruvate. Studies with normal gingival fibroblasts also noted that pyruvate lessened the antiproliferative toxicity of caffeic acid, EGCG, TFdiG, black tea theaflavin mixture, and G. biloba extract [55]. Caspase-mediated cleavage of poly(ADP-ribose) polymerase (PARP), an indicator of apoptosis, occurred upon exposure of HSC-2 cells to 200 and 250 μg/mL pomegranate extract. In cells cotreated with pomegranate extract and pyruvate, PARP cleavage was greatly reduced, indicating that pomegranate extract-induced apoptosis was a function of the induction of oxidative stress [14].

Of the various mediators of pro-oxidant polyphenol cytotoxicity, pyruvate may have the most far reaching effects, as it is a component in some, but not in all, commercially available formulations of cell culture media. Furthermore, even within one particular type of medium, there may be different formulations, some with and some without pyruvate. Thus, RPMI, McCoy, Medium 199, and MEM typically lack pyruvate, whereas MEM contains pyruvate. DMEM, the most commonly used cell culture medium, has various formulations, the majority of which contain pyruvate and others, much lesser in number, lack pyruvate. Figure 6 shows the comparative cytotoxicities to HSC-2 cells of pomegranate extract and olive fruit extract in two commercially available formulations of DMEM, one with and the other without pyruvate. The mediating effect of pyruvate on H_2O_2 toxicity presumably is not well known, as the pyruvate content of the cell culture medium is seldom reported in the experimental design. Thus, cell responses to a pro-oxidant polyphenol may vary among laboratories, depending on the pyruvate content of the exposure media, which should be noted in publications.

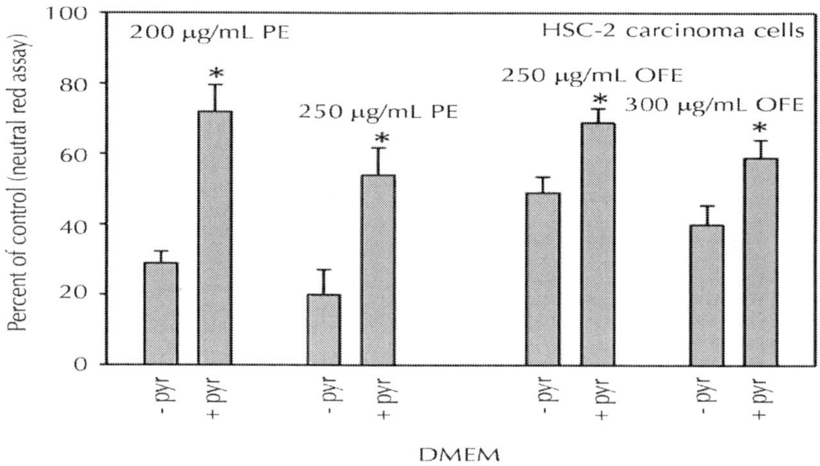

Figure 6: Comparative toxicities of pomegranate extract (PE) and olive fruit extract (OFE) towards human oral HSC-2 carcinoma cells, in Dulbecco's modified Eagle medium (DMEM) commercially formulated without pyruvate (−pyr) and with pyruvate (+pyr). Cell proliferation was quantified with the neutral red assay, a cell viability assay. Data are presented as the mean percent of control ± S.E.M. Data for pomegranate extract are from Weisburg et al. [14] and for olive fruit extract from Schuck (unpublished).

In addition to directly scavenging exogenous H_2O_2, pyruvate has other properties to mediate oxidative stress. Pyruvate readily penetrates cells by a specific H^+-monocarboxylic cotransporter [56] and is a readily oxidized fuel, enhancing the cytosolic energy state to maintain cellular functioning in the face of metabolic challenge [57]. Pyruvate dampens the mitochondrial generation of ROS, stabilizes mitochondrial ATP production compromised by oxidative stress, maintains the mitochondrial membrane potential under oxidative stress [58], and upregulates the expression of glutathione peroxidase, which is concerned with mitochondrial scavenging of H_2O_2 and in maintaining GSH levels [59]. The pyruvate content of the cell culture medium is an amendment that should be reckoned in studies of pro-oxidant polyphenols.

Cobalt

Divalent cobalt (Co^{2+}) catalyzes the decomposition of H_2O_2 to H_2O and O_2, without an accompanying change in its valence [60]. Additions of mixtures of black tea polyphenols or of green tea polyphenols to cell culture medium generated significant amounts of H_2O_2, which progressively decreased in the presence of increasing concentrations of added Co^{2+}, as $CoCl_2$. As a scavenger of H_2O_2, amendments of cell culture medium with Co^{2+} afforded protection to transformed oral S-G cells from green and black tea polyphenol mixtures [61]. The antiproliferative effects of EGCG to oral HCS-2 and HSG carcinoma cells [49] and to normal gingival fibroblasts [62], of a black tea theaflavin mixture to transformed gingival GT1 fibroblasts and tongue carcinoma CAL27 cells [35], and of Gingko biloba extract [13] and of pomegranate extract [14] to HSC-2 cells were substantially lessened in the presence of $CoCl_2$. Activation of caspase-3, an indicator of apoptosis, was noted in HSC-2 cells treated with pomegranate extract; when cotreated with pomegranate extract and $CoCl_2$, activation of caspase-3 was greatly lessened [14]. In many of these studies [13, 14, 35], Co^{2+} afforded almost complete protection against polyphenol-induced growth inhibition (Figure 7). At the concentrations used in these studies, Co^{2+} usually provided greater protection from oxidative damage than catalase.

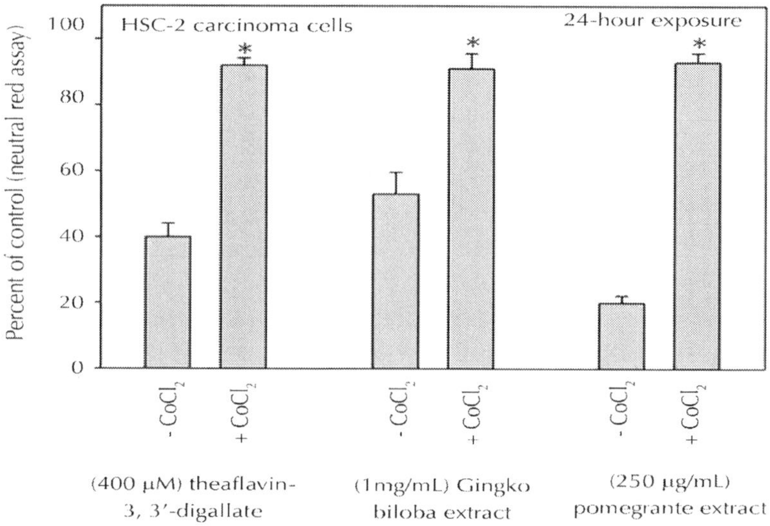

Figure 7: Comparative toxicities of theaflavin-3,3′-digallate, Ginkgo biloba extract, and pomegranate extract towards human oral HSC-2 carcinoma cells, untreated (control) and treated with cobalt chloride. Cell proliferation was quantified with the neutral red assay, a cell viability assay. Data are presented as the mean percent of control ± S.E.M. Data for the theaflavin-3,3′-gallate are from Schuck et al. [19], for G. biloba extract from Babich et al. [13], and for pomegranate extract from Weisburg et al. [14].

Iron

Nakagawa et al. [63] noted that o-phenanthroline, a Fe^{2+}-chelating agent, suppressed EGCG-induced cell death in cultured osteoclastic cells. In a cell-free system, it was further shown that the reduction of Fe^{3+} to Fe^{2+} by EGCG triggered a Fenton reaction to form a highly reactive hydroxyl radical from the EGCG-generated H_2O_2 [$H_2O_2 + Fe^{2+} \rightarrow OH^{\cdot} + OH^{-} + Fe^{3+}$]. Subsequent studies by Nakagawa et al. [23] noted that the growth inhibitory and apoptotic-inducing effects of EGCG towards Jurkat cells were partially suppressed by o-phenanthroline. These investigators suggested that the mode of cytotoxicity of EGCG was through the generation of H_2O_2, which triggered the Fe^{2+}-dependent Fenton reaction, thereby generating highly toxic $OH^{\cdot}$ radicals to inhibit growth and to induce apoptotic cell death.

Lipid peroxidation in transformed oral S-G cells exposed to protocatechuic acid was enhanced in the presence of Fe^{2+} [37]. Similarly, Fe^{2+} potentiated the EGCG-lipid peroxidation towards oral cancerous and normal gingival fibroblasts [12]. In both studies, the polyphenol per se had little effect on lipid peroxidation, thus, indicating the involvement of a Fenton reaction.

Copper

Several studies have shown that plant polyphenols in the presence of metal ions cause oxidative damage to DNA. Divalent copper (Cu^{2+}), one of the most redox sensitive metal ions in cells, is closely associated with chromatin. In a study by Bhat et al. [42], freshly isolated peripheral lymphocytes were treated with caffeic acid, and DNA damage was evaluated with the comet assay. DNA breakage, observed with 200 μM caffeic acid, was progressively lessened in the presence of increasing concentrations of neocuproine, a Cu^{1+}-specific chelating agent. Apparently, DNA breakage by caffeic acid involved endogenous copper, with Cu^{1+} an intermediate in the mechanistic pathway leading to DNA cleavage. Treatments with catalase and SOD reduced caffeic acid-mediated DNA damage; also, caffeic acid-mediated lipid peroxidation was lessened in the presence of neocuproine. Apparently, both Cu^{1+} and ROS were involved in oxidative damage by caffeic acid. To explain the mechanism for DNA damage by caffeic acid, it was postulated that the polyphenol bound both DNA and Cu^{2+} to form a ternary complex, leading to the reduction of Cu^{2+} to Cu^{1+}, which subsequently interacted with polyphenol-generated H_2O_2 to afford $OH^{\cdot}$ radicals via a Fenton reaction. Similarly, damage of cellular DNA in lymphocytes was noted upon exposures to EGCG, gallic acid [42], and quercetin [48], with neocuproine sequestering DNA damage. Others have noted interactions between pro-oxidant polyphenols, ROS, and $Cu^{2+} \rightarrow Cu^{1+}$ to cause oxidative DNA damage although other mechanistic pathways, not involving a ternary complex but rather direct interactions between the polyphenol and Cu^{2+}, were suggested [38, 64, 65].

CELLULAR RESPONSES AS MEDIATED BY PRO-OXIDANT POLYPHENOL CONCENTRATION

Three distinct cellular responses appear to result from exposure to polyphenols, with each response dependent upon the concentration and pro-oxidant nature of the polyphenols. (a) A mild exposure causes mild oxidative stress and thereby ignites cellular antioxidant defense systems. (b) An intermediate to high exposure gradually overwhelms the antioxidant defense systems and induces apoptotic cell death. (c) A very high exposure quickly overwhelms the cellular antioxidant defenses and causes oxidative damage leading to cell death by necrosis. In designing experiments to identify oxidative stress as the causative mechanism of cytotoxicity, a careful balance is needed between the generated ROS and the experimental variable (i.e., scavenger or potentiator of ROS toxicity) mediating oxidative stress.

As noted, glutathione, the major contributor for maintaining the redox state of the cell, exists in both a reduced (GSH) and an oxidized form (GSSG). Maintaining suitable levels of GSH is crucial to counteract oxidative stress and involves the transactivation of phase II detoxification/antioxidant genes encoding enzymes for GSH synthesis. Several gene response elements are involved in transcriptional regulation of GSH metabolism, including the antioxidant/electrophile response elements (AREs/EpREs), which have promoters with a specific consensus sequence of nucleotides that respond to molecules with antioxidant properties. Thus, plant-derived polyphenols, as antioxidants, regulate AREs/EpREs, leading to the synthesis of GSH. Although the specific mechanism is unclear, one approach suggested that the pro-oxidant nature of the polyphenol was the critical factor, in that auto-oxidation of the polyphenol generated ROS, which lessened the concentration of GSH, thereby to ignite transcriptional activation of -glutamylcysteine synthetase [5]. Exposure of COS-1 cells, an immortalized African green monkey kidney cell line, to noncytotoxic concentrations of quercetin enhanced synthesis of GSH through upregulation of -glutamylcysteine synthetase. Exposure of these cells to high concentrations of polyphenols led to elevated levels of ROS, which quickly depleted GSH stores and thereby increased cellular susceptibility to oxidative free radical attack, resulting in cell death by either apoptosis and/or necrosis [66].

Studies with gingival fibroblasts showed that a 4-hour exposure to a nontoxic concentration of TFdiG [19] or of EGCG [12] stimulated the resynthesis of GSH, often to levels exceeding baseline. Early studies by Arrick et al. [67] noted that after a brief exposure to GSH depleters, cells rapidly resynthesized GSH, often overshooting normal levels. It has been suggested that a potential health benefit for the consumption of polyphenols at dietary levels is to generate low levels of ROS, so as to induce mild oxidative stress and thereby boost antioxidant defense systems to counteract potential challenge by elevated levels of ROS, perhaps through mitochondrial respiration [5, 9].

Raza and John [68], using PC cells, derived from a pheochromocytoma of the rat adrenal medulla, showed that the level of EGCG was critical in evoking either a defensive mechanism or cell death by apoptosis. Low levels of EGCG (e.g., 50 μM) apparently induced mild oxidative stress, igniting cellular antioxidant defenses, such as the stimulation of GSH synthesis and increased activity of glutathione-S-transferase. A higher level EGCG (e.g., 400 μM) potentiated oxidative stress, as indicated by a persistent elevated intracellular level of ROS that overwhelmed and masked the antioxidant defensive mechanisms, leading to disruption of the intracellular GSH pool and an increase in lipid peroxidation. Also noted at 400 μM EGCG was an increased expression of the enzyme, cytochrome P450 2E1, a member of the cytochrome P450 mixed-function oxidase system involved in the metabolism of ROS and a participant in cellular oxidative stress-related toxicity.

In an interesting study by Hsuuw and Chan [69] with human breast cancer MCF-7 cells, responses were compared after exposures to moderate (20–50 μM) and to high levels (100–400 μM) of EGCG. Moderate levels of EGCG-induced apoptosis and elevated levels of EGCG induced cell necrosis (cell lysis). At moderate levels of EGCG, cell viability was lessened, and apoptosis was induced, and both correlated with increased oxidative stress, as indicated by intracellular generation of ROS, a loss of mitochondrial membrane potential, activation of caspase-3, caspase-9, and c-Jun-terminal kinase (JNK), and an increased expression level of Bax protein and a decreased level of Bcl-2 protein, thereby shifting the Bax-Bcl-2 ratio to favor apoptosis. Pretreatment of the MCF-7 cells with the antioxidants, NAC or -tocopherol, attenuated intracellular ROS generation and rescued the cells from apoptotic death. Exposure to 25–50 μM EGCG did not adversely affect ATP production; pretreatment with antimycin A, an inhibitor of mitochondrial respiratory activity, rescued the cells from apoptotic death, indicating that high levels of ATP were required to induce apoptosis. Conversely, at the high concentrations of EGCG, lesser levels of intracellular ROS and of apoptotic cells and only minor effects on caspase and JNK activations, on the Bax-Bcl-2 ratio, and on mitochondrial membrane potential were noted. Instead, cell death correlated with leakage of LDH, a sign of necrosis. Lowered ATP levels were observed at the high levels of EGCG. It was suggested that the switch from apoptotic death to necrosis was controlled by the intracellular level of ATP, with high ATP levels favoring apoptosis and decreased levels favoring necrosis.

STUDIES ON THE PRO-OXIDANT EFFECTS OF POLYPHENOLS IN LABORATORY ANIMALS AND HUMANS

The limited in vitro studies on the pro-oxidant nature of natural photochemical dwarfs the minuscule research with laboratory animal model systems. Li et al. [70] cited their study as "the first demonstration that EGCG induces ROS formation and consequently causes DNA oxidative damage in tumor cells in animals." NCr nu/nu mice bearing human lung cancer H1299 xenograft tumors were maintained on a diet supplemented with 0.1, 0.3, and 0.5% EGCG. At the end of the 45-day experimental period, tumor growth was found to be dose dependently inhibited by EGCG. To correlate tumor inhibition with EGCG-induced oxidative stress, immunohistochemistry staining of the xenograft tumors was performed. The parameters evaluated included (a) the formation of the oxidative DNA-product, 8-hydroxyl-2′-deoxyguanosine (8-OHdG), a commonly used marker of oxidative stress, (b) the formation of phosphorylated histone 2A variant X (-H2AX), a cellular marker for the presence of double-strand DNA breaks, which can be caused by ROS, and (c) apoptotic activity, as measured by the induction of caspase-3.

Dose-dependent induction of all three biochemical parameters was observed in the xenograft tumors.

Animal studies using elevated doses of EGCG have reported hepatotoxicity linked to oxidative damage. Treatment of CF mice with a single high oral dose of 1,500 mg/kg EGCG or two once-daily doses of 750 mg/kg EGCG caused liver necrosis, associated with induction of apoptosis and increased lipid peroxidation and -histone 2AX protein expression. Plasma levels of 8-isoprostane, a nonenzymatic marker of arachidonic acid oxidation, were also increased upon EGCG treatment [71].

Apparently, there are no other laboratory animal studies on the pro-oxidant effects of polyphenol nutraceuticals.

It has been noted that oral consumption of tea protected against carcinomas in the human oral cavity. Li et al. [72] showed that oral and topical administration of a mixed green tea preparation significantly reduced the size of oral precancerous lesions and the incidence of micronucleated oral mucosa cells in leukoplakia patients. Halder et al. [73] noted similar results in patients with oral leukoplakia who were administered a regimen of black tea. These studies did not evaluate whether the pro-oxidative nature of the teas was involved in chemoprevention. However, studies by Yang et al. [74], Lee et al. [75], and Lambert et al. [76] observed that holding tea solutions in the oral cavity or chewing tea leaves generated high salivary levels of catechins and theaflavins, accompanied with high levels of salivary H_2O_2. Catechin- and theaflavin-generated salivary H_2O_2 may have played a role in controlling the precancerous lesions in the studies by Li et al. [72] and Halder et al. [73].

CONCLUDING REMARKS

Cell response to a polyphenol challenge reflects a duality of toxicities, that of the polyphenol per se and that of the generated ROS, both of which modulate chemical signaling pathways leading to antiproliferative and apoptotic effects. To distinguish between the two chemical challenges, studies in the absence and in the presence of ROS scavengers are warranted. Some of the research strategies employed to elucidate oxidative stress are well established, for example, the use of glutathione depleters and ROS scavenging enzymes; other strategies are less known, that is, use of divalent cobalt and pyruvate to scavenge ROS. In studies in our laboratory, cobalt (usually, at 250 μM $CoCl_2$) was a more efficient scavenger of H_2O_2 than was catalase (usually, at 100 Units/mL). Most important for studies of oxidative stress is acknowledgement of the ROS scavenging property of pyruvate, as it is incorporated in formulations of some commercially available media and not in others. DMEM is probably the most utilized cell culture medium, yet its pyruvate is seldom noted in the experimental design. Of the numerous formulations of DMEM, relatively few lack pyruvate. As noted in our laboratory [55], the magnitude of the cellular response to a polyphenol differed significantly if the study was performed in DMEM without pyruvate as compared to using DMEM with pyruvate. As the pyruvate status of the medium is seldom incorporated into the description of the experimental design, it can be presumed that the ROS scavenging property of pyruvate is not well known.

ACKNOWLEDGMENT

Appreciation is expressed to Stern College for Women of Yeshiva University for kindly providing funding for the research reviewed in this paper.

REFERENCES

1. R. Béliveau and D. Gingras, "Role of nutrition in preventing cancer," Canadian Family Physician, vol. 53, no. 11, pp. 1905–1911, 2007.
2. S. Nair, W. Li, and A. N. T. Kong, "Natural dietary anti-cancer chemopreventive compounds: redox-mediated differential signaling mechanisms in cytoprotection of normal cells versus cytotoxicity in tumor cells," Acta Pharmacologica Sinica, vol. 28, no. 4, pp. 459–472, 2007.
3. H. Pelicano, D. Carney, and P. Huang, "ROS stress in cancer cells and therapeutic implications," Drug Resistance Updates, vol. 7, no. 2, pp. 97–110, 2004.

4. H. U. Simon, A. Haj-Yehia, and F. Levi-Schaffer, "Role of reactive oxygen species (ROS) in apoptosis induction," Apoptosis, vol. 5, no. 5, pp. 415–418, 2000.
5. J. O. Moskaug, H. Carlsen, M. C. Myhrstad, and R. Blomhoff, "Polyphenols and glutathione synthesis regulation," The American Journal of Clinical Nutrition, vol. 71, supplement, pp. 16985–17025, 2005.
6. E. A. Decker, "Phenolics: prooxidants or antioxidants?" Nutrition Reviews, vol. 55, no. 11, pp. 396–398, 1997.
7. L. R. Fukumoto and G. Mazza, "Assessing antioxidant and prooxidant activities of phenolic compounds," Journal of Agricultural and Food Chemistry, vol. 48, no. 8, pp. 3597–3604, 2000.
8. R. Vittal, Z. E. Selvanayagam, Y. Sun et al., "Gene expression changes induced by green tea polyphenol (−)-epigallocatechin-3-gallate in human bronchial epithelial 21BES cells analyzed by DNA microarray," Molecular Cancer Therapeutics, vol. 3, no. 9, pp. 1091–1099, 2004.
9. J. D. Lambert and R. J. Elias, "The antioxidant and pro-oxidant activities of green tea polyphenols: a role in cancer prevention," Archives of Biochemistry and Biophysics, vol. 501, no. 1, pp. 65–72, 2010.
10. C. S. Yang, J. D. Lambert, Z. Hou, J. Ju, G. Lu, and X. Hao, "Molecular targets for the cancer preventive activity of tea polyphenols," Molecular Carcinogenesis, vol. 45, no. 6, pp. 431–435, 2006.
11. P. C. Chai, L. H. Long, and B. Halliwell, "Contribution of hydrogen peroxide to the cytotoxicity of green tea and red wines," Biochemical and Biophysical Research Communications, vol. 304, no. 4, pp. 650–654, 2003.
12. J. H. Weisburg, D. B. Weissman, T. Sedaghat, and H. Babich, "In vitro cytotoxicity of epigallocatechin gallate and tea extracts to cancerous and normal cells from the human oral cavity," Basic and Clinical Pharmacology and Toxicology, vol. 95, no. 4, pp. 191–200, 2004.
13. H. Babich, N. J. Ackerman, F. Burekhovich, H. L. Zuckerbraun, and A. G. Schuck, "Gingko biloba leaf extract induces oxidative

stress in carcinoma HSC-2 cells," Toxicology in Vitro, vol. 23, no. 6, pp. 992–999, 2009.

14. J. H. Weisburg, A. G. Schuck, M. S. Silverman et al., "Pomegranate extract, A prooxidant with antiproliferative and proapoptotic activities preferentially towards carcinoma cells," Anti-Cancer Agents in Medicinal Chemistry, vol. 10, no. 8, pp. 634–643, 2010.
15. T. Lapidot, M. D. Walker, and J. Kanner, "Can apple antioxidants inhibit tumor cell proliferation? Generation of H_2O_2 during interaction of phenolic compounds with cell culture media," Journal of Agricultural and Food Chemistry, vol. 50, no. 11, pp. 3156–3160, 2002.
16. L. H. Long, M. V. Clement, and B. Halliwell, "Artifacts in cell culture: rapid generation of hydrogen peroxide on addition of (−)-epigallocatechin, (−)-epigallocatechin gallate, (+)-catechin, and quercetin to commonly used cell culture media," Biochemical and Biophysical Research Communications, vol. 273, no. 1, pp. 50–53, 2000.
17. H. Babich, M. E. Krupka, H. A. Nissim, and H. L. Zuckerbraun, "Differential in vitro cytotoxicity of (−)-epicatechin gallate (ECG) to cancer and normal cells from the human oral cavity," Toxicology in Vitro, vol. 19, no. 2, pp. 231–242, 2005.
18. H. Babich, H. L. Zuckerbraun, and S. M. Weinerman, "In vitro cytotoxicity of (−)-catechin gallate, a minor polyphenol in green tea," Toxicology Letters, vol. 171, no. 3, pp. 171–180, 2007.
19. A. G. Schuck, M. B. Ausubel, H. L. Zuckerbraun, and H. Babich, "Theaflavin-3,3′-digallate, a component of black tea: an inducer of oxidative stress and apoptosis," Toxicology in Vitro, vol. 22, no. 3, pp. 598–609, 2008.
20. H. Babich, R. T. Gottesman, E. J. Liebling, and A. G. Schuck, "Theaflavin-3-gallate and theaflavin-3′-gallate, polyphenols in black tea with prooxidant properties," Basic and Clinical Pharmacology and Toxicology, vol. 103, no. 1, pp. 66–74, 2008.
21. T. Lapidot, M. D. Walker, and J. Kanner, "Antioxidant and prooxidant effects of phenolics on pancreatic β-cells in vitro," Journal of Agricultural and Food Chemistry, vol. 50, no. 25, pp. 7220–7225, 2002.

22. K. W. Lee, H. J. Hur, H. J. Lee, and C. Y. Lee, "Antiproliferative effects of dietary phenolic substances and hydrogen peroxide," Journal of Agricultural and Food Chemistry, vol. 53, no. 6, pp. 1990–1995, 2005.

23. H. Nakagawa, K. Hasumi, J. T. Woo, K. Nagai, and M. Wachi, "Generation of hydrogen peroxide primarily contributes to the induction of Fe(II)-dependent apoptosis in Jurkat cells by (−)-epigallocatechin gallate," Carcinogenesis, vol. 25, no. 9, pp. 1567–1574, 2004.

24. S. G. Rhee, T. S. Chang, W. Jeong, and D. Kang, "Methods for detection and measurement of hydrogen peroxide inside and outside of cells," Molecules and Cells, vol. 29, no. 6, pp. 539–549, 2010.

25. G. Y. Yang, J. Liao, K. Kim, E. J. Yurkow, and C. S. Yang, "Inhibition of growth and induction of apoptosis in human cancer cell lines by tea polyphenols," Carcinogenesis, vol. 19, no. 4, pp. 611–616, 1998.

26. T. Nakazato, K. Ito, Y. Miyakawa et al., "Catechin, a green tea component, rapidly induces apoptosis of myeloid leukemic cells via modulation of reactive oxygen species production in vitro and inhibits tumor growth in vivo," Haematologica, vol. 90, no. 3, pp. 317–325, 2005.

27. T. Nakazato, K. Ito, Y. Ikeda, and M. Kizaki, "Green tea component, catechin, induces apoptosis of human malignant B cells via production of reactive oxygen species," Clinical Cancer Research, vol. 11, no. 16, pp. 6040–6049, 2005.

28. T. Yamamoto, J. Lewis, J. Wataha et al., "Roles of catalase and hydrogen peroxide in green tea polyphenol-induced chemopreventive effects," Journal of Pharmacology and Experimental Therapeutics, vol. 308, no. 1, pp. 317–323, 2004.

29. G. Y. Yang, J. Liao, C. Li et al., "Effect of black and green tea polyphenols on c-jun phosphorylation and H_2O_2 production in transformed and non-transformed human bronchial cell lines: possible mechanisms of cell growth inhibition and apoptosis induction," Carcinogenesis, vol. 21, no. 11, pp. 2035–2039, 2000.

30. M. M. Chan, K. J. Soprano, K. Weinstein, and D. Fong, "Epigallocatechin-3-gallate delivers hydrogen peroxide to

induce death of ovarian cancer cells and enhances their cisplatin susceptibility," Journal of Cellular Physiology, vol. 207, no. 2, pp. 389–396, 2006.

31. T. Atsumi, K. Tonosaki, and S. Fujisawa, "Induction of early apoptosis and ROS-generation activity in human gingival fibroblasts (HGF) and human submandibular gland carcinoma (HSG) cells treated with curcumin," Archives of Oral Biology, vol. 51, no. 10, pp. 913–921, 2006.
32. S. Yeu, C. M. Sun, H. H. Chuang, and P. T. Chang, "Studies on the cytotoxic mechanisms of ginkgetin in a human ovarian adenocarcinoma cell line," Naunyn-Schmiedeberg's, vol. 362, no. 1, pp. 82–90, 2000.
33. J. M. Feugang, F. Ye, D. Y. Zhang et al., "Cactus pear extracts induce reactive oxygen species production and apoptosis in ovarian cancer cells," Nutrition and Cancer, vol. 62, no. 5, pp. 692–699, 2010.
34. H. Babich, A. R. Selevan, and E. R. Ravkin, "Glutathione as a mediator of the in vitro cytotoxicity of a green tea polyphenol extract," Toxicology Mechanisms and Methods, vol. 17, no. 6, pp. 357–369, 2007.
35. H. Babich, S. M. Pinsky, E. T. Muskin, and H. L. Zuckerbraun, "In vitro cytotoxicity of a theaflavin mixture from black tea to malignant, immortalized, and normal cells from the human oral cavity,"Toxicology in Vitro, vol. 20, no. 5, pp. 677–688, 2006.
36. J. Lu, C. T. Ho, G. Ghai, and . Kuan Yu Chen, "Differential effects of theaflavin monogallates on cell growth, apoptosis, and Cox-2 gene expression in cancerous versus normal cells," Cancer Research, vol. 60, no. 22, pp. 6465–6471, 2000.
37. H. Babich, A. Sedletcaia, and B. Kenigsberg, "In vitro cytotoxicity of protocatechuic acid to cultured human cells from oral tissue: involvement in oxidative stress," Pharmacology and Toxicology, vol. 91, no. 5, pp. 245–253, 2002.
38. A. Furukawa, S. Oikawa, M. Murata, Y. Hiraku, and S. Kawanishi, "(−)-Epigallocatechin gallate causes oxidative damage to isolated and cellular DNA," Biochemical Pharmacology, vol. 66, no. 9, pp. 1769–1778, 2003.

39. H. Babich, A. Stern, and E. Borenfreund, "Eugenol cytotoxicity evaluated with continuous cell lines,"Toxicology in Vitro, vol. 7, no. 2, pp. 105–109, 1993.

40. B. R. You and W. H. Park, "Gallic acid-induced lung cancer cell death is related to glutathione depletion as well as reactive oxygen species increase," Toxicology in Vitro, vol. 24, no. 5, pp. 1356–1362, 2010.

41. S. C. Sahu, D. I. Ruggles, and M. W. O'Donnell, "Prooxidant activity and toxicity of nordihydroguaiaretic acid in clone-9 rat hepatocyte cultures," Food and Chemical Toxicology, vol. 44, no. 10, pp. 1751–1757, 2006.

42. S. H. Bhat, A. S. Azmi, and S. M. Hadi, "Prooxidant DNA breakage induced by caffeic acid in human peripheral lymphocytes: involvement of endogenous copper and a putative mechanism for anticancer properties," Toxicology and Applied Pharmacology, vol. 218, no. 3, pp. 249–255, 2007.

43. S. B. Sarikçioğlu, G. Öner, and E. Tercan, "Antioxidant effect of EGb 761 on hydrogen peroxide-induced lipoperoxidation of G-6-PD deficient erythrocytes," Phytotherapy Research, vol. 18, no. 10, pp. 837–840, 2004.

44. H. Babich, A. G. Reisbaum, and H. L. Zuckerbraun, "In vitro response of human gingival epithelial S-G cells to resveratrol," Toxicology Letters, vol. 114, no. 1–3, pp. 143–153, 2000.

45. L. Elbling, R. M. Weiss, O. Teufelhofer et al., "Green tea extract and (−)-epigallocatechin-3-gallate, the major tea catechin, exert oxidant but lack antioxidant activities," FASEB Journal, vol. 19, no. 7, pp. 807–809, 2005.

46. Z. Hou, S. Sang, H. You et al., "Mechanism of action of (−)-epigallocatechin-3-gallate: auto-oxidation- dependent inactivation of epidermal growth factor receptor and direct effects on growth inhibition in human esophageal cancer KYSE 150 cells," Cancer Research, vol. 65, no. 17, pp. 8049–8056, 2005.

47. G. C. Yen, J. W. Ju, and C. H. Wu, "Modulation of tea and tea polyphenols on benzo(a)pyrene-induced DNA damage in Chang liver cells," Free Radical Research, vol. 38, no. 2, pp. 193–200, 2004.

48. S. M. Hadi, S. H. Bhat, A. S. Azmi, S. Hanif, U. Shamim, and M. F. Ullah, "Oxidative breakage of cellular DNA by plant polyphenols: a putative mechanism for anticancer properties," Seminars in Cancer Biology, vol. 17, no. 5, pp. 370–376, 2007.
49. A. Ishino, S. Mita, S. Watanabe, and H. Sakagami, "Effect of anticancer drugs, metals and antioxidants on cytotoxic activity of epigallocatechin gallate," Anticancer Research, vol. 19, no. 5, pp. 4343–4348, 1999.
50. S. Fujisawa, T. Atsumi, M. Ishihara, and Y. Kadoma, "Cytotoxicity, ROS-generation activity and radical-scavenging activity of curcumin and related compounds," Anticancer Research, vol. 24, no. 2, pp. 563–569, 2004.
51. C. Chen, G. Shen, V. Hebbar, R. Hu, E. D. Owuor, and A. N. T. Kong, "Epigallocatechin-3-gallate-induced stress signals in HT-29 human colon adenocarcinoma cells," Carcinogenesis, vol. 24, no. 8, pp. 1369–1378, 2003.
52. K. A. Nath, H. Enright, L. Nutter, M. Fischereder, J. N. Zou, and R. P. Hebbel, "Effect of pyruvate on oxidant injury to isolated and cellular DNA," Kidney International, vol. 45, no. 1, pp. 166–176, 1994.
53. R. T. Mallet, "Pyruvate: metabolic protector of cardiac performance," Proceedings of the Society for Experimental Biology and Medicine, vol. 223, no. 2, pp. 136–148, 2000.
54. A. Shostak, L. Gotloib, R. Kushnier, and V. Wajsbrot, "Protective effect of pyruvate upon cultured mesothelial cells exposed to 2 mM hydrogen peroxide," Nephron, vol. 84, no. 4, pp. 362–366, 2000.
55. H. Babich, E. J. Liebling, R. F. Burger, H. L. Zuckerbraun, and A. G. Schuck, "Choice of DMEM, formulated with or without pyruvate, plays an important role in assessing the in vitro cytotoxicity of oxidants and prooxidant nutraceuticals," In vitro Cellular & Developmental Biology, vol. 45, no. 5-6, pp. 226–233, 2009.
56. C. K. Garcia, J. L. Goldstein, R. K. Pathak, R. G. W. Anderson, and M. S. Brown, "Molecular characterization of a membrane transporter for lactate, pyruvate, and other monocarboxylates: implications for the Cori cycle," Cell, vol. 76, no. 5, pp. 865–873, 1994.

57. R. Bunger, R. T. Mallet, and D. A. Hartman, "Pyruvate-enhanced phosphorylation potential and inotropism in normoxic and postischemic isolated working heart. Near-complete prevention of reperfusion contractile failure," European Journal of Biochemistry, vol. 180, no. 1, pp. 221–233, 1989.
58. X. Wang, E. Perez, R. Liu, L. J. Yan, R. T. Mallet, and S. H. Yang, "Pyruvate protects mitochondria from oxidative stress in human neuroblastoma SK-N-SH cells," Brain Research, vol. 1132, no. 1, pp. 1–9, 2007.
59. F. M. Maiorino, R. Brigelius-Flohé, K. D. Aumann, A. Roveri, D. Schomburg, and L. Flohé, "[5] Diversity of glutathione peroxidases," Methods in Enzymology, vol. 252, pp. 38–48, 1995.
60. M. K. Eberhardt, C. Santos, and M. A. Soto, "Formation of hydroxyl radicals and Co^{3+} in the reaction of Co^{2+}-EDTA with hydrogen peroxide. Catalytic effect of Fe^{3+}," Biochimica et Biophysica Acta, vol. 1157, no. 1, pp. 102–106, 1993.
61. H. Babich, T. Gold, and R. Gold, "Mediation of the in vitro cytotoxicity of green and black tea polyphenols by cobalt chloride," Toxicology Letters, vol. 155, no. 1, pp. 195–205, 2005.
62. T. Sakagami, K. Satoh, M. Ishihara et al., "Effect of cobalt ion on radical intensity and cytotoxic activity of antioxidants," Anticancer Research, vol. 20, no. 5, pp. 3143–3150, 2000.
63. H. Nakagawa, M. Wachi, J. T. Woo et al., "Fenton reaction is primarily involved in a mechanism of (3)-epigallocatechin-3-gallate to induce osteoclastic cell death," Biochemical and Biophysical Research Communications, vol. 292, no. 1, pp. 94–101, 2002.
64. S. Inoue, K. Ito, K. Yamamoto, and S. Kawanishi, "Caffeic acid causes metal-dependent damage to cellular and isolated DNA through H_2O_2 formation," Carcinogenesis, vol. 13, no. 9, pp. 1497–1502, 1992.
65. S. Oikawa, A. Furukawa, H. Asada, K. Hirakawa, and S. Kawanishi, "Catechins induce oxidative damage to cellular and isolated DNA through the generation of reactive oxygen species," Free Radical Research, vol. 37, no. 8, pp. 881–890, 2003.

66. M. C. W. Myhrstad, H. Carlsen, O. Nordström, R. Blomhoff, and J. Ø. Moskaug, "Flavonoids increase the intracellular glutathione level by transactivation of the γ-glutamylcysteine synthetase catalytical subunit promoter," Free Radical Biology and Medicine, vol. 32, no. 5, pp. 386–393, 2002.
67. B. A. Arrick, C. F. Nathan, O. W. Griffith, and Z. A. Cohn, "Glutathione depletion sensitizes tumor cells to oxidative cytolysis," The Journal of Biological Chemistry, vol. 257, no. 3, pp. 1231–1237, 1982.
68. H. Raza and A. John, "Green tea polyphenol epigallocatechin-3-gallate differentially modulates oxidative stress in PC12 cell compartments," Toxicology and Applied Pharmacology, vol. 207, no. 3, pp. 212–220, 2005.
69. Y. D. Hsuuw and W. H. Chan, "Epigallocatechin gallate dose-dependently induces apoptosis or necrosis in human MCF-7 cells," Annals of the New York Academy of Sciences, vol. 1095, pp. 428–440, 2007.
70. G. X. Li, Y. K. Chen, Z. Hou et al., "Pro-oxidative activities and dose-response relationship of (−)-epigallocatechin-3-gallate in the inhibition of lung cancer cell growth: a comparative study in vivo and in vitro," Carcinogenesis, vol. 31, no. 5, pp. 902–910, 2010.
71. J. D. Lambert, M. J. Kennett, S. Sang, K. R. Reuhl, J. Ju, and C. S. Yang, "Hepatotoxicity of high oral dose (−)-epigallocatechin-3-gallate in mice," Food and Chemical Toxicology, vol. 48, no. 1, pp. 409–416, 2010.
72. N. Li, Z. Sun, C. Han, and J. Chen, "The chemopreventive effects of tea on human oral precancerous mucosa lesions," Proceedings of the Society for Experimental Biology and Medicine, vol. 220, no. 4, pp. 218–224, 1999.
73. A. Halder, R. Raychowdhury, A. Ghosh, and M. De, "Black tea (Camellia sinensis) as a chemopreventive agent in oral precancerous lesions," Journal of Environmental Pathology, Toxicology and Oncology, vol. 24, no. 2, pp. 141–144, 2005.
74. C. S. Yang, M. J. Lee, and L. Chen, "Human salivary tea catechin levels and catechin esterase activities: implication in human cancer prevention studies," Cancer Epidemiology Biomarkers and Prevention, vol. 8, no. 1, pp. 83–89, 1999.

75. M. J. Lee, J. D. Lambert, S. Prabhu et al., "Delivery of tea polyphenols to the oral cavity by green tea leaves and black tea extract," Cancer Epidemiology Biomarkers and Prevention, vol. 13, no. 1, pp. 132–137, 2004.
76. J. D. Lambert, S. J. Kwon, J. Hong, and C. S. Yang, "Salivary hydrogen peroxide produced by holding or chewing green tea in the oral cavity," Free Radical Research, vol. 41, no. 7, pp. 850–853, 2007.

Chapter 4

Nutraceutical Interventions for Promoting Healthy Aging in Invertebrate Models

Yuqing Dong[1,2], Sujay Guha[1], Xiaoping Sun[3], Min Cao[1,2], Xiaoxia Wang[1], and Sige Zou[3]

[1]Department of Biological Sciences, Clemson University, 132 Long Hall, Clemson, SC 29634, USA

[2]Institute for Engaged Aging, Clemson University, 535A Edwards Hall, Clemson, SC 29634, USA

[3]Laboratory of Experimental Gerontology, National Institute on Aging, 251 Bayview Boulevard, Baltimore, MD 21224, USA

ABSTRACT

Aging is a complex and inevitable biological process that is associated with numerous chronically debilitating health effects. Development of effective interventions for promoting healthy aging is an active but challenging area of research. Mechanistic studies in various model organisms, noticeably two invertebrates,Caenorhabditis elegans and

Drosophila melanogaster, have identified many genes and pathways as well as dietary interventions that modulate lifespan and healthspan. These studies have shed light on some of the mechanisms involved in aging processes and provide valuable guidance for developing efficacious aging interventions. Nutraceuticals made from various plants contain a significant amount of phytochemicals with diverse biological activities. Phytochemicals can modulate many signaling pathways that exert numerous health benefits, such as reducing cancer incidence and inflammation, and promoting healthy aging. In this paper, we outline the current progress in aging intervention studies using nutraceuticals from an evolutionary perspective in invertebrate models.

INTRODUCTION

Aging is associated with a gradual decline of physiological and cognitive functions [1]. Age is a major risk factor for numerous diseases, such as diabetes, cancer, and various degenerative diseases, including Alzheimer's, Huntington's, and Parkinson's diseases [2, 3]. A major challenge to health care systems around the world is how to encourage and maintain a healthy lifespan in large and increasing populations of elderly individuals [1]. Significant progress has been made over the past two decades in elucidating the molecular mechanisms of aging [4–7]. Hundreds of genetic factors, called longevity-related genes, have been identified to modulate lifespan and healthspan in model organisms ranging from yeast, worms, flies, and rodents. A number of the longevity-related genes fall into three conserved nutrient sensing pathways: target-of-rapamycin (TOR), insulin/IGF-1-like signaling (IIS), and sirtuin pathways [7–9]. These pathways primarily sense cellular amino acid, glucose, and NAD^+ or NAD^+/NADH levels, respectively, (Figure 1). Genome-wide association studies have linked some longevity-related genes discovered in model organisms to human longevity, such as insulin-like growth factor receptor (IGF1R) and forkhead transcription factor (FOXO3a) in the IIS pathway [10–14]. Mechanistic studies of aging have led to the proposal of a number of hypotheses of aging. Perhaps the most prominent is the free radical hypothesis of aging, which states that free radicals, such as reactive oxygen species, generated from metabolism inflict oxidative damage to macromolecules, including protein, DNA, and lipid. Accumulation of such oxidative damage with age causes biological aging and eventually results in death [15, 16]. However, numerous studies have now shown that this simplified version of free radical hypothesis of aging is not necessarily sufficient to explain the mechanisms underlying aging processes. Another is the hormesis hypothesis of aging, which has been frequently used to interpret the prolongevity effects induced by nutraceuticals. Hormesis theory states that mild stress-induced stimulation of defense response at the organismal level results in biologically beneficial effects and extends lifespan and healthspan. The defense response involves many protective mechanisms and influences with gene expression and metabolism [16–18]. Therefore, significant efforts in developing interventions for promoting healthy aging have been devoted to identify effective ways to modulate metabolism and stress [8].

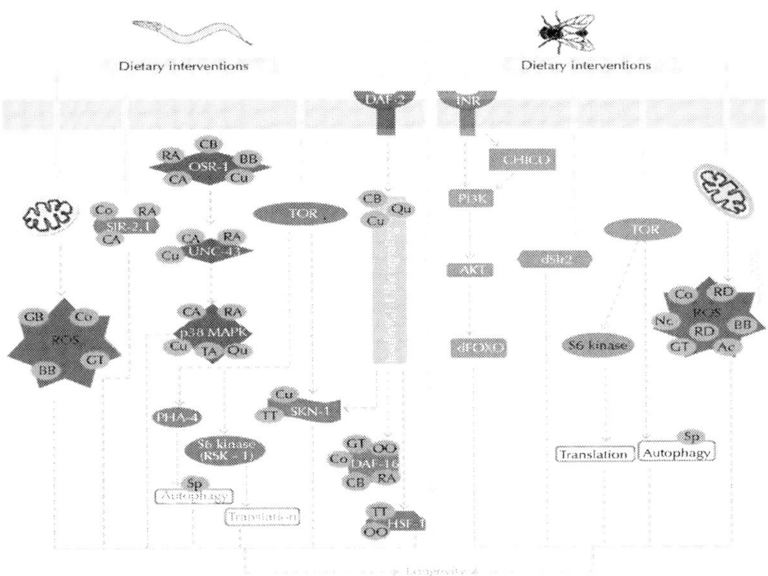

Figure 1: Nutrient sensing signaling and stress response pathways in C. elegans and D. melanogaster. Dietary interventions promote healthy lifespan and stress resistance through at least three conserved nutrient sensing

pathways, IIS, TOR, and sirtuins. Many nutraceuticals improve healthspan by regulating ROS/redox state, some are known to interact with the IIS and osmotic resistant pathways to extend lifespan, and some modulate lifespan extension via autophagy mediated by TOR signaling. Each nutraceutical is represented by a green circle and the location of the green circle indicates the possible target of a nutraceutical action*. Ac: açai, BB: blueberry, CA: caffeic acid, CB: cranberry, Co: cocoa, Cu: curcumin, GB: ginkgo biloba, GT: green tea, Ne: nectarine, OO: olive oil, Qu: quercetin, RD: rosa damascene, RA: rosmarinic acid, Sp: spermidine, TA: tannic acid, and TT: thioflavin T. *ROS presented in the figure indicates the regulation of ROS/redox state.

Dietary restriction (DR) by modest reduction of all macronutrients or protein content is one of the most effective environmental interventions for promoting healthy aging [19–25]. However, it is challenging to implement dietary restriction in humans. A number of genes, including components in the conserved nutrient sensing pathways, have been identified to mediate lifespan extension induced by DR [9, 26]. As an alternative to DR, many studies have been devoted to the identification of pharmacological and nutraceutical reagents that extend lifespan and healthspan, some of which may mimic the effect of DR [27, 28]. Due to their short lifespan and rich genetic resource, Drosophila melanogaster and Caenorhabditis elegans, have been in the forefront of research on aging interventions. In this paper, we focus on aging interventions in invertebrate models using nutraceuticals, which refer to extracts made out of edible fruits, vegetables, and herbs. Interventions with pharmacological reagents, such as resveratrol, metformin, and rapamycin, have been discussed in numerous comprehensive reviews elsewhere [28–33], and will not be described in this paper. Here we attempt to integrate the findings from studies on the mechanisms of aging and nutraceutical interventions with an evolutionary perspective.

DIETARY BOTANICALS AND PHYTOCHEMICALS

Numerous studies have demonstrated the effects of nutraceuticals from fruit or plant extracts in reducing oxidative damage and promoting healthy aging in invertebrate models. The active ingredients in nutraceuticals that are generally produced by plants as "secondary compounds" appear to help plants overcome stressful conditions

[34–36]. The beneficial properties of nutraceuticals can be attributed to the varieties of phytochemicals, such as flavonoids, anthocyanin glycosides, triterpenoids, and proanthocyanidin oligomers [37–39]. In this section, we will describe the antiaging properties of several nutraceuticals made from fruits, spices, and teas, which are commonly consumed by humans in daily life.

Ginkgo Biloba Extract (EGB 761)

Ginkgo is widely used as an ancient Chinese medicine due to its benefits to the health of the elderly [40]. The ginkgo biloba extract has been proposed to help prevent Alzheimer's disease (AD). The herbal extract EGb761 made from Ginkgo biloba is rich in flavonoids and terpenoids, which improves the functions of platelet and nerve cells and the blood flow to the nervous system and brain, probably due to its antioxidant properties [41, 42]. In wild type worms, EGb761 extends lifespan by ~10%, reduces accumulation of oxidative damage, and represses expression of thermal stress-response protein hsp-16.2 [43–46]. In addition, EGb 761 inhibits Aβ oligomerization and alleviates Aβ-induced paralysis in the C. elegans model of AD [47]. Aβoligomerization is a hallmark of AD [48]. However, it remains to be determined whether ginkgo biloba can promote longevity in other models, such as Drosophila. Nevertheless, combined with studies demonstrating the neuroprotective function of EGb 761 in mammalian cells [49, 50], findings in worms suggest that the beneficial effect of EGb761 in prevention and treatment of AD could at least partially be due to its capacity to resist oxidative stress and reduce detrimental protein aggregation.

Blueberry

Blueberry contains a wide array of polyphenols and offers a host of health benefits [51]. Supplementation of blueberry preserves learning and memory in aged F344 rats by improving neuronal function [52]. Polyphenols from blueberry significantly increase the lifespan of C. elegans [53]. In C. elegans, blueberry supplementation decreases age-related accumulation of the intracellular level of lipofuscin, a biomarker for age-related cellular damage, and reduces the level of 4-Hydroxynonenal level, a biomarker for lipid peroxidation. In addition, blueberry polyphenols improve the pharyngeal pumping rates of the aged worms and increase thermotolerance, suggesting that blueberry improves worms' healthspan. Mechanistic studies indicate that blueberry supplementation reduces mRNA levels of heat shock proteins and requires OSR-1/UNC-43/SEK-1, components of the osmotic stress resistance pathway [53], to promote longevity in C. elegans. A recent study has demonstrated that blueberry extracts extends mean lifespan by approximately 10% in Drosophila [54]. Associated with lifespan extension, blueberry increased expression of several antioxidant genes, including superoxide dismutase 1 (SOD1), SOD2, and catalase (CAT) in Drosophila. Further genetic studies are needed to establish whether any of the antioxidant genes are required for lifespan extension induced by blueberry in Drosophila. These findings in invertebrates suggest that antioxidant machinery and/or osmotic stress pathway may play a pivotal and common role in mediating lifespan extension induced by blueberry.

Oregano and Cranberry

Cranberry and oregano possess multiple medicinal properties, such as antimicrobial, antiviral, antimutagenic, antiangiogenic, and antioxidation functions [55, 56]. A mixture of oregano and cranberry (OC) extract increased lifespan in Mexican fruit flies (mexfly) in a diet composition dependent manner [57]. OC extended lifespan in mexflies fed diets with a relatively higher sugar to protein ratio, but not in mexflies fed diets with a lower sugar to protein ratio. OC supplementation did not compromise the lifetime reproductive output, a parameter of healthspan. In addition, OC supplementation in middle age was sufficient to promote longevity [58]. However, lifespan was not increased when OC was supplemented only in young age or old age. These findings point out the importance of considering diet composition and implementing time in developing an efficacious aging intervention.

We have recently assessed the effect of cranberry extract alone on lifespan and healthspan in C. elegans. Our data indicate that the cranberry extract alone is sufficient to prolong lifespan in C. elegans [59]. The cranberry-mediated lifespan extension was suppressed almost completely by the absence of DAF-16, a forkhead transcription factor in the IIS. Cranberry supplementation did not further extend the lifespan of DAF-2 or AGE-1 mutants, either. DAF-2 and AGE-1 are insulin-like receptor and phosphatidylinositol 3 kinase (PI3 K) in the IIS pathway of C. elegans [60, 61]. Cranberry is high in antioxidants and phytochemicals, including proanthocyanidins and vitamin C, which may neutralize free radicals and reduce oxidative damage, and more importantly modulating signaling transduction pathways [51, 62]. Our findings suggest that IIS and DAF-16 play an important role in cranberry mediated prolongevity in C. elegans. Further studies are warranted to determine to what extent the prolongevity effect of cranberry attributes to its antioxidant properties.

Nectarine and Açai

Nectarine is a globally consumed fruit [63], and açai is a fruit indigenous to the Amazon River area [64]. Both fruits contain various kinds of bioactive phytochemicals [2, 3, 65]. Nectarine supplementation can extend lifespan in flies fed standard, DR, and high-fat diets. The lifespan extension induced by nectarine is associated with increased lifetime reproductive output and reduced lipid oxidation [66]. In contrast, supplementation of açai pulp promotes survival in Drosophila fed a high-fat but not a standard diet [67]. The diet composition dependent effect of açai is also evident in the mexfly. Açai supplementation promotes the survival of the mexfly fed a high-fat and high-sugar diet but no other nonfat diets [68]. Along with the OC study described above, the importance of diet composition is also evident in aging intervention studies using pharmacological agents, such as resveratrol. Studies in Drosophila, mexfly, and mice have shown that the prolongevity effects of resveratrol depend on diet composition [69–71]. These studies again stress the importance of diet composition in modulating the health benefits of nutraceuticals. Moreover, both nectarine and açai can promote the survival of flies with sod1 deficiency. sod1 deficient flies have a short lifespan and experience high levels of oxidative damage. These findings suggest that nectarine and açai possess antioxidant activities at the organismal level [26, 67, 72]. However, it remains to be determined the mechanisms underlying the prosurvival effects of nectarine and açai.

Rosa Damascena

A hybrid rose species, Rosa damascena, is traditionally used to make rose oil and water in cosmetic and food industries [73, 74]. Extracts from R. damascena contains numerous volatile organic compounds including various terpenes such as citronellol, heneicosane, and disiloxane, and also polyphenols, such as quercetin, myricetin, kaempferol, and gallic acid [75]. R. damascena extracts have been shown to possess biological properties that are protective against microbial infection, seizures in rats, and toxicity of amyloid beta in neurons, a biomarker of AD [76–79]. An R. damascena extract has been found to increase both mean and maximum lifespan in Drosophila. This extract also enhances flies' resistance to oxidative stress and low iron stress, but paradoxically increases flies' sensitivity to heat stress. Molecular studies indicate that the rose extract reduces the heat-induced expression of a major heat shock protein HSP70 and a small mitochondrial heat shock protein HSP22 [80]. Therefore, it has been proposed that R. damascena extract extends lifespan by protecting flies against iron-induced stress.

Cocoa Polyphenols

Numerous polyphenols with high antioxidant activities, such as flavonoids, have been isolated from cocoa [81]. Flavonoid-enriched cocoa powder reduces oxidative stress in C. elegans [82]. Moreover, the flavonoid-enriched cocoa powder also significantly increases the lifespan of wild type N2 worms [82]. This lifespan extension was dependent on both SIR-2.1 and DAF-16. In addition, the flavonoid-enriched cocoa powder failed to exert any longevity effect on daf-2 mutants, indicating that cocoa powder-mediated lifespan extension may also act through the IIS. Considering the mechanism by which SIR-2.1 and IIS modulates lifespan, cocoa polyphenols may promote longevity by reducing oxidative stress, influencing metabolism, and altering chromatin structure.

Green Tea

Green tea contains polyphenolic catechins that have been reported to have a number of health benefits, including prevention of AD,

Parkinson's disease, and heart disease [83]. Green tea can protect against angiogenesis and tumor formation [84]. The health benefits of green tea are mainly attributed to bioactive properties of its phytochemical constituents [85, 86]. A number of polyphenolic catechins, including epicatechin (EC), epicatechin-3-gallate (ECG), epigallocatechin (EGC), epigallocatechin-3-gallate (EGCG), catechin, and gallocatechin (GC), are abundant in green tea [83]. Among these, EGCG is perhaps the most abundant catechin in green tea, and has been reported to induce antioxidant enzymes, including glutathione peroxidase, catalase, and glutathione S-transferase, in mice [87]. Similar results were obtained in worms and flies. Supplementation with green tea catechins increases the antioxidant enzymatic activity of superoxide dismutase and catalase in Drosophila [88, 89]. Green tea extends median lifespan by 36% and mean lifespan by 16% in flies [89]. In C. elegans, EGCG treatment significantly reduces ROS levels under both normal culture and oxidative stress conditions, and increases the expression of sod-3 and hsp-16.2 [90, 91]. Administration of EGCG significantly extends the lifespan of C. elegans under heat stress and oxidative stress [90, 91]. In addition, a recent study demonstrated that L-theanine, a unique amino acid particularly enriched in green tea, promoted the survival of C. elegans in the presence of paraquat [92]. L-theanine has been reported to provide broad health benefits, such as antitumor, AD prevention, and blood pressure reduction [93–96]. Together, these findings suggest that green tea increases lifespan and stress resistance partially through its antioxidant properties.

Olive Oil

Numerous studies have demonstrated that consumption of olive oil has multiple beneficial effects on health and longevity in humans [97–101]. The abundance of phenolic compounds present in olive oil is considered to play an important role in exerting these healthy effects. Lastest studies have revealed that the tyrosol, one of the most abundant phenols in olive oil [102], significantly promotes the longevity of C. elegans, as well as the resistance to thermal and oxidative stress [103]. IIS/DAF-16 and HSF-1 are required in tyrosol-mediated prolongevity in C. elegans. hsp-12.6, a coregulated target gene of DAF-16, and HSF-1 [104], is significantly upregulated in adult worms after the tyrosol treatment. It has been reported that small heat shock proteins, including HSP-12.6, can extend lifespan and delay polyglutamine protein aggregation in C. elegans [104]. HSF-1 is critical for maintaining protein homeostasis. Together, these studies suggest that tyrosol in olive oil extends lifespan by increasing oxidative stress resistance and thermotolerance as well as improving protein homeostasis [103].

Quercetin and Tannic Acid

Quercetin (3,3′,4′,5,7-pentahydroxyflavone) is one of the most important dietary flavonoids present in a wide array of foods. Numerous health benefits are linked to consumption of fruits and vegetables containing a high content of quercetin [105]. Quercetin has been shown to dramatically increase the lifespan of C. elegans[106–109]. However, the studies on the molecular basis yielded conflicting results. One study indicates that lifespan extension induced by quercetin supplementation requires DAF-2 and AGE-1, the components of C. elegans insulin/IGF signaling [106]. Quecertin induces nuclear translocation of DAF-16 [110]. This study suggests that the IIS pathway mediates the prolongevity effect of quercetin. However, this hypothesis has been challenged by another study, which indicates that quercetin mediated lifespan extension does not depend on DAF-16 [107]. Additional studies are warranted to resolve the conflicting observation and understand the mechanisms underlying the prolongevity effect of quercetin.

Tannic acid (TA) belongs to tannins, which are secondary metabolites of plants with many health effects. TA has the biological activities to prevent neurodegeneration [111], pathogen infection [112, 113], carcinogenesis [114, 115], and oxidative damage [116, 117]. In addition, TA significantly increases the lifespan of worms [108]. The IIS pathway and DAF-16 are not essential for the prolongevity effects of TA. Only SEK-1, one component of the p38 MAPK pathway, has been shown to be essential for TA-induced lifespan extension.

Transcriptome studies indicate that quercetin affects expression of genes in the TGF-β signaling, insulin-like signaling, and p38 MAPK pathways, while TA changes expressions of genes in the TGF-β and the p38 MAPK pathways as well as the amino acid metabolism. Together, these studies suggest that TGF-β and p38 MAPK pathways play crucial roles in mediating the prolongevity effects of quercetin and TA [118].

Caffeic Acid and Rosmarinic Acid

Caffeic acid (3-(3,4-dihydroxyphenyl)-2-propenoic acid; CA) and Rosmarinic acid (α-o-caffeoyl-3, 4-dihydroxyphenyl lactic acid; RA) are abundantly present in a variety of fruits, vegetables, and herbs. CA and RA have anticarcinogenesis, antioxidant, antimicrobial, anti-inflammatory, and antirheumatic properties [119–123]. CA and RA can prolong the healthy lifespan of C. elegans [124]. DAF-16, SIR-2.1, OSR-1, UNK-43, and SEK-1 are required for RA triggered lifespan extension. Similar results have been obtained for CA mediated lifespan extension except that DAF-16 does not appear to play a pivotal role in this lifespan extension. Together, these findings suggest that CA and RA promote lifespan extension through overlapping pathways involved in metabolism and stress response.

Spermidine

Spermidine is an important polyamine presented in citrus fruit and soybean, and has effects on epigenetic modifications, autophagy, and necrosis [125, 126]. Polyamine concentrations and autophagy have been shown to decline in various organisms, including humans [127, 128]. Supplementation of spermidine prolongs the lifespan of C. elegans and Drossssophila by 15% and 30%, respectively [125]. A growing number of evidence shows that autophagy plays an essential role in the regulation of lifespan mediated by the TOR pathway [129–131]. Autophagy is required for spermidine-mediated lifespan extension in both worms and flies. It will be interesting to see if spermidine and its derivatives can confer lifespan extension in humans by, at least in part, enhancing the autophagy.

Curcumin and Thioflavin T

Curcumin (diferuloylmethane) is the pharmacologically active substance in turmeric (Curcuma longa), and has been widely used as a herbal medicine in Asia. A large body of evidence indicates that curcumin possesses many biological activities, such as antioxidative, anti-inflammatory, anticancer, chemopreventive, and antineurodegenerative properties [132–136]. With its pleiotropic activities, curcumin has been considered as a potential aging intervention compound. Studies in Drosophila and C. elegans have demonstrated that curcumin can delay aging and prolong the lifespan [137, 138]. Curcumin-treated flies exhibited enhanced resistance against to oxidative stress, improved locomotor activity, and higher tolerance to chemotherapy drugs. Curcumin reduces expression of Methuselah, a longevity gene, and genes in the IIS and JNK pathways. Curcumin-induced lifespan extension has been shown to be mediated by the OSR-1/UNC-43/SEK-1 pathway and possible the members of the IIS and SKN-1 pathways in C. elegans.

Protein homeostasis is an essential lifespan modulator of animals [139]. Dysfunctional protein homeostasis results in protein misfolding and aggregation, which is a hallmark of aging and age-related diseases [140]. ThT is a compound known to bind and inhibit aggregation of amyloids, which leads to various neurodegenerative disorders [141]. Flavonoid thioflavin T (4-(3,6-dimethyl-1,3-benzothiazole-3-ium-2-yl)-N,N-dimethylaniline chloride, ThT) has been shown to promote longevity in C. elegans [142]. Treatment with 50 μM ThT, a concentration conferring a largest increase of lifespan, significantly suppresses the protein aggregation mediated paralysis in worms. The prolongevity effect of ThT depends on HSF-1 and SKN-1, but independent of the IIS. Importantly, the beneficial effects of ThT require an intact machinery for maintaining protein homeostasis. Like ThT, curcumin also has the ability to bind and inhibit protein aggregates [142]. Treatment of both ThT and curcumin does not yield synergistic effects on longevity. ThT and curcumin may act through similar mechanisms, for example, improving protein homeostasis, to modulate aging and age-related disorders.

CONCLUDING REMARKS

Nutraceuticals made from widely-consumed plant products represent both opportunities and challenges in aging interventions. On the one hand, nutraceuticals from plant extracts can promote longevity and improve healthspan through multiple mechanisms, such as reducing oxidative stress, altering signaling pathways, influencing metabolism, and maintaining protein homeostasis (Figure 1). Nutraceuticals that promote longevity in invertebrate models are often made from edible and relatively easily accessible fruits, vegetables, spices, and other plant products. Thus, when appropriately implemented nutraceutical interventions can be efficient and cost-effective ways for promoting healthy aging in humans.

On the other hand, major challenges remain to be addressed in implementing nutraceutical interventions. First, nutraceuticals contain numerous bioactive phytochemicals. It is a daunting task to determine which phytochemicals are the active ingredients for promoting health benefits, or determine how the beneficial effects are mediated through synergistic actions among multiple ingredients. Second, more research is needed to elucidate the molecular mechanisms underlying the actions of nutraceuticals, in order to better assess the effects of nutraceutical supplementation on the health of animals. For instance, more recent studies have revealed that ROS plays a physiologically vital role in signal transduction, gene regulation, and redox regulation [143]. The previous concept of the "free radical hypothesis of aging" has been modified and diversified ROS a messenger role in addition to its toxicity. Thus, it would be harmful to eliminate ROS completely. In the context, although many nutraceuticals have antioxidant properties, we need more careful exams to unravel the molecular basis rather than simply attribute to their ability to scavenge free radicals. It is in line with this notion that some nutraceuticals, such as quercetin and blueberry polyphenols, modulate lifespan by activating molecular pathways independent to their direct antioxidant capacity [53, 107]. Third, it is essential to investigate the impact of both short-term and long-term nutraceutical interventions on healthspan. A recently developed behavioral monitoring system capable of recording lifetime behavioral changes at a high resolution represents an example of a useful tool for assessing healthspan in invertebrates [144]. Lastly, an important and challenging issue is to further understand how the effects of nutraceuticals depend on diet composition and genetic background. The prolongevity effects of several nutraceuticals have been shown to depend on the timing of supplementation and the composition of diet [143, 145–148]. The diet composition-dependent effects will have a significant impact on the increasing demand for personalized nutritional intervention. The individual and synergistic effects of nutraceuticals as a component of dietary composition will require further study and scientific scrutiny, particularly since little regulatory oversight regarding their sale for human consumption currently exists.

We have highlighted how nutraceuticals prevent and protect against aging and stress in invertebrate models. Some of nutraceuticals and their synthetic derivatives are being tested for their therapeutic potential [149]. Numerous promising results have been obtained in model organisms that suggest evolutionarily conserved mechanisms are involved in their beneficial effects. Much progress has been made to decipher the molecular mechanisms of aging shared among multiple species, which provide valuable guidance for aging interventions. However, further extensive studies will be required to demonstrate whether any nutraceuticals or pharmaceuticals can effectively delay aging or age-related disease in humans. A wider range of additional assays should be considered to help us better understand aging processes and improve the quality and quantity of human life in the foreseeable future.

ACKNOWLEDGMENTS

The authors thank Edward Spangler and Robert Kosinski for critical reading of the paper. They also thank Cole Murbach for editorial assistance. This work was supported by the Creative Inquiry fund at Clemson University and the Yamada Research Grant to Y. Dong and M. Cao, an American Federation of Aging Research (AFAR) grant to Y. Dong and the Intramural Research Program at the National Institute on Aging, NIH to S. Zou.

REFERENCES

1. E. De Luca d'Alessandro, S. Bonacci, and G. Giraldi, "Aging populations: the health and quality of life of the elderly," La Clinica Terapeutica, vol. 162, no. 1, pp. e13–e18, 2011.
2. V. Pizza, A. Agresta, C. W. ĎAcunto, M. Festa, and A. Capasso, "Neuroinflamm-aging and neurodegenerative diseases: an overview," CNS and Neurological Disorders, vol. 10, no. 5, pp. 621–634, 2011.
3. L. Fontana, "Modulating human aging and age-associated diseases," Biochimica et Biophysica Acta, vol. 1790, no. 10, pp. 1133–1138, 2009.

4. F. Peysselon and S. Ricard-Blum, "Understanding the biology of aging with interaction networks,"Maturitas, vol. 69, no. 2, pp. 126–130, 2011.
5. G. M. Martin, "The biology of aging: 1985–2010 and beyond," The FASEB Journal, vol. 25, no. 11, pp. 3756–3762, 2011.
6. C. J. Kenyon, "The genetics of ageing," Nature, vol. 464, no. 7288, pp. 504–512, 2010.
7. L. Fontana, L. Partridge, and V. D. Longo, "Extending healthy life span-from yeast to humans," Science, vol. 328, no. 5976, pp. 321–326, 2010.
8. M. C. Haigis and B. A. Yankner, "The aging stress response," Molecular Cell, vol. 40, no. 2, pp. 333–344, 2010.
9. N. Alic and L. Partridge, "Death and dessert: nutrient signalling pathways and ageing," Current Opinion in Cell Biology, vol. 23, no. 6, pp. 738–743, 2011.
10. E. Ziv and D. Hu, "Genetic variation in insulin/IGF-1 signaling pathways and longevity," Ageing Research Reviews, vol. 10, no. 2, pp. 201–204, 2011.
11. M. Soerensen, S. Dato, K. Christensen et al., "Replication of an association of variation in the FOXO3Agene with human longevity using both case-control and longitudinal data," Aging Cell, vol. 9, no. 6, pp. 1010–1017, 2010.
12. L. Pawlikowska, D. Hu, S. Huntsman et al., "Association of common genetic variation in the insulin/IGF1 signaling pathway with human longevity," Aging Cell, vol. 8, no. 4, pp. 460–472, 2009.
13. F. Flachsbart, A. Caliebe, R. Kleindorp, et al., "Association of FOXO3A variation with human longevity confirmed in German centenarians," Proceedings of the National Academy of Sciences of the United States of America, vol. 106, no. 8, pp. 2700–2705, 2009.
14. B. J. Willcox, T. A. Donlon, Q. He et al., "FOXO3A genotype is strongly associated with human longevity," Proceedings of the National Academy of Sciences of the United States of America, vol. 105, no. 37, pp. 13987–13992, 2008.

15. D. Harman, "Aging: a theory based on free radical and radiation chemistry," Journal of Gerontology, vol. 11, no. 3, pp. 298–300, 1956.
16. A. B. Salmon, A. Richardson, and V. I. Pérez, "Update on the oxidative stress theory of aging: does oxidative stress play a role in aging or healthy aging?" Free Radical Biology and Medicine, vol. 48, no. 5, pp. 642–655, 2010.
17. M. Ristow and K. Zarse, "How increased oxidative stress promotes longevity and metabolic health: the concept of mitochondrial hormesis (mitohormesis)," Experimental Gerontology, vol. 45, no. 6, pp. 410–418, 2010.
18. S. I. S. Rattan, H. Sejersen, R. A. Fernandes, and W. Luo, "Stress-mediated hormetic modulation of aging, wound healing, and angiogenesis in human cells," Annals of the New York Academy of Sciences, vol. 1119, no. 1, pp. 112–121, 2007.
19. C. M. McCay, M. F. Crowell, and L. A. Maynard, "The effect of retarded growth upon the length of life span and upon the ultimate body size," Journal of Nutrition, vol. 10, pp. 63–79, 1935.
20. R. Weindruch, R. L. Walford, S. Fligiel, and D. Guthrie, "The retardation of aging in mice by dietary restriction: longevity, cancer, immunity and lifetime energy intake," Journal of Nutrition, vol. 116, no. 4, pp. 641–654, 1986.
21. J. C. Jiang, E. Jaruga, M. V. Repnevskaya, and S. M. Jazwinski, "An intervention resembling caloric restriction prolongs life span and retards aging in yeast," The FASEB Journal, vol. 14, no. 14, pp. 2135–2137, 2000.
22. D. J. Clancy, D. Gems, E. Hafen, S. J. Leevers, and L. Partridge, "Dietary restriction in long-lived dwarf flies," Science, vol. 296, no. 5566, p. 319, 2002.
23. W. Mair and A. Dillin, "Aging and survival: the genetics of life span extension by dietary restriction,"Annual Review of Biochemistry, vol. 77, pp. 727–754, 2008.
24. R. J. Colman, R. M. Anderson, S. C. Johnson et al., "Caloric restriction delays disease onset and mortality in rhesus monkeys," Science, vol. 325, no. 5937, pp. 201–204, 2009.
25. E. L. Greer and A. Brunet, "Different dietary restriction regimens extend lifespan by both independent and overlapping genetic

pathways in C. elegans," Aging Cell, vol. 8, no. 2, pp. 113–127, 2009.

26. M. D. W. Piper and A. Bartke, "Diet and aging," Cell Metabolism, vol. 8, no. 2, pp. 99–104, 2008.
27. E. Morselli, M. C. Maiuri, M. Markaki et al., "Caloric restriction and resveratrol promote longevity through the Sirtuin-1-dependent induction of autophagy," Cell Death and Disease, vol. 1, no. 1, article e10, 2010.
28. L. Mouchiroud, L. Molin, N. Dallière, and F. Solari, "Life span extension by resveratrol, rapamycin, and metformin: the promise of dietary restriction mimetics for an healthy aging," BioFactors, vol. 36, no. 5, pp. 377–382, 2010.
29. A. R. Mendelsohn and J. W. Larrick, "Rapamycin as an antiaging therapeutic? Targeting mammalian target of rapamycin to treat Hutchinson-Gilford progeria and neurodegenerative diseases,"Rejuvenation Research, vol. 14, no. 4, pp. 437–441, 2011.
30. G. E. Mullin, "Red wine, grapes, and better health—resveratrol," Nutrition in Clinical Practice, vol. 26, no. 6, pp. 722–723, 2011.
31. O. Vang, N. Ahmad, C. A. Baile et al., "What is new for an old molecule? systematic review and recommendations on the use of resveratrol," PLoS ONE, vol. 6, no. 6, Article ID e19881, 2011.
32. S. Bulterijs, "Metformin as a geroprotector," Rejuvenation Research, vol. 14, no. 5, pp. 469–482, 2011.·
33. F. Madeo, T. Eisenberg, S. Büttner, C. Ruckenstuhl, and G. Kroemer, "Spermidine: a novel autophagy inducer and longevity elixir," Autophagy, vol. 6, no. 1, pp. 160–162, 2010.
34. B. Winkel-Shirley, "Biosynthesis of flavonoids and effects of stress," Current Opinion in Plant Biology, vol. 5, no. 3, pp. 218–223, 2002.
35. P. Reymond, H. Weber, M. Damond, and E. E. Farmer, "Differential gene expression in response to mechanical wounding and insect feeding in Arabidopsis," Plant Cell, vol. 12, no. 5, pp. 707–719, 2000.·
36. D. Hermsmeier, U. Schittko, and I. T. Baldwin, "Molecular interactions between the specialist herbivore Manduca sexta (Lepidoptera, Sphingidae) and its natural host Nicotiana attenuata. I. Large-scale changes in the accumulation of growth-

and defense-related plant mRNAs," Plant Physiology, vol. 125, no. 2, pp. 683–700, 2001.

37. R. Tsao, "Chemistry and biochemistry of dietary polyphenols," Nutrients, vol. 2, no. 12, pp. 1231–1246, 2010.

38. D. O. Kennedy and E. L. Wightman, "Herbal extracts and phytochemicals: plant secondary metabolites and the enhancement of human brain function," Advances in Nutrition, vol. 2, no. 1, pp. 32–50, 2011.

39. A. Salminen, A. Kauppinen, and K. Kaarniranta, "Phytochemicals suppress nuclear factor-kappaB signaling: impact on health span and the aging process," Current Opinion in Clinical Nutrition & Metabolic Care, vol. 15, no. 1, pp. 23–28, 2012. ·

40. T. Y. Wu, C. P. Chen, and T. R. Jinn, "Traditional Chinese medicines and Alzheimer's disease,"Taiwanese Journal of Obstetrics and Gynecology, vol. 50, no. 2, pp. 131–135, 2011.

41. B. A. Abdel-Wahab and S. M. Abd El-Aziz, "Ginkgo biloba protects against intermittent hypoxia-induced memory deficits and hippocampal DNA damage in rats," Phytomedicine, vol. 19, no. 5, pp. 444–450, 2012.

42. F. V. DeFeudis and K. Drieu, "Ginkgo biloba extract (EGb 761) and CNS functions: basic studies and clinical applications," Current Drug Targets, vol. 1, no. 1, pp. 25–58, 2000.

43. A. Kampkötter, T. Pielarski, R. Rohrig et al., "The Ginkgo biloba extract EGb761 reduces stress sensitivity, ROS accumulation and expression of catalase and glutathione S-transferase 4 inCaenorhabditis elegans," Pharmacological Research, vol. 55, no. 2, pp. 139–147, 2007.

44. J. Vining Smith and Y. Luo, "Elevation of oxidative free radicals in Alzheimer's disease models can be attenuated by Ginkgo biloba extract EGb 761," Journal of Alzheimer's Disease, vol. 5, no. 4, pp. 287–300, 2003.

45. A. Strayer, Z. Wu, Y. Christen, C. D. Link, and Y. Luo, "Expression of the small heat-shock protein Hsp16-2 in Caenorhabditis elegans is suppressed by Ginkgo biloba extract EGb 761," The FASEB Journal, vol. 17, no. 15, pp. 2305–2307, 2003.

46. Z. Wu, J. V. Smith, V. Paramasivam et al., "Ginkgo biloba extract EGb 761 increases stress resistance and extends life span of

Caenorhabditis elegans," Cellular and Molecular Biology, vol. 48, no. 6, pp. 725–731, 2002.

47. Y. Wu, Z. Wu, P. Butko et al., "Amyloid-β-induced pathological behaviors are suppressed by Ginkgo biloba extract EGB 761 and ginkgolides in transgenic Caenorhabditis elegans," Journal of Neuroscience, vol. 26, no. 50, pp. 13102–13113, 2006.

48. V. H. Finder and R. Glockshuber, "Amyloid-β aggregation," Neurodegenerative Diseases, vol. 4, no. 1, pp. 13–27, 2007.

49. Y. Luo, J. V. Smith, V. Paramasivam et al., "Inhibition of amyloid-β aggregation and caspase-3 activation by the Ginkgo biloba extract EGb761," Proceedings of the National Academy of Sciences of the United States of America, vol. 99, no. 19, pp. 12197–12202, 2002.

50. J. V. Smith, A. J. Burdick, P. Golik, I. Khan, D. Wallace, and Y. Luo, "Anti-apoptotic properties of Ginkgo biloba extract EGb 761 in differentiated PC12 cells," Cellular and Molecular Biology, vol. 48, no. 6, pp. 699–707, 2002.

51. C. C. Neto, "Cranberry and blueberry: evidence for protective effects against cancer and vascular diseases," Molecular Nutrition and Food Research, vol. 51, no. 6, pp. 652–664, 2007.

52. J. A. Joseph, B. Shukitt-Hale, N. A. Denisova et al., "Reversals of age-related declines in neuronal signal transduction, cognitive, and motor behavioral deficits with blueberry, spinach, or strawberry dietary supplementation," Journal of Neuroscience, vol. 19, no. 18, pp. 8114–8121, 1999.

53. M. A. Wilson, B. Shukitt-Hale, W. Kalt, D. K. Ingram, J. A. Joseph, and C. A. Wolkow, "Blueberry polyphenols increase lifespan and thermotolerance in Caenorhabditis elegans," Aging Cell, vol. 5, no. 1, pp. 59–68, 2006.

54. C. Peng, Y. Zuo, K. M. Kwan et al., "Blueberry extract prolongs lifespan of Drosophila melanogaster,"Experimental Gerontology, vol. 47, no. 2, pp. 170–178, 2012.

55. E. Pappas and K. M. Schaich, "Phytochemicals of cranberries and cranberry products: characterization, potential health effects, and processing stability," Critical Reviews in Food Science and Nutrition, vol. 49, no. 9, pp. 741–781, 2009.

56. F. E. Babili, J. Bouajila, J. P. Souchard et al., "Oregano: chemical analysis and evaluation of its antimalarial, antioxidant, and cytotoxic activities," Journal of Food Science, vol. 76, no. 3, pp. C512–C518, 2011.

57. S. Zou, J. R. Carey, P. Liedo, D. K. Ingram, B. Yu, and R. Ghaedian, "Prolongevity effects of an oregano and cranberry extract are diet dependent in the mexican fruit fly (Anastrepha ludens)," Journals of Gerontology, vol. 65, no. 1, pp. 41–50, 2010.

58. S. Zou, J. R. Carey, P. Liedo, D. K. Ingram, and B. Yu, "Prolongevity effects of a botanical with oregano and cranberry extracts in Mexican fruit flies: examining interactions of diet restriction and age," Age, vol. 34, no. 2, pp. 269–279, 2012.

59. S. Guha, M. Cao, R. M. Kane, A. M. Savino, S. Zou, and Y. Dong, "The longevity effect of cranberry extract in Caenorhabditis elegans is modulated by daf-16 and osr-1," Age. In press.

60. E. Cohen and A. Dillin, "The insulin paradox: aging, proteotoxicity and neurodegeneration," Nature Reviews Neuroscience, vol. 9, no. 10, pp. 759–767, 2008.

61. J. Z. Morris, H. A. Tissenbaum, and G. Ruvkun, "A phosphatidylinositol-3-OH kinase family member regulating longevity and diapause in Caenorhabditis elegans," Nature, vol. 382, no. 6591, pp. 536–539, 1996.

62. X. He and H. L. Rui, "Cranberry phytochemicals: isolation, structure elucidation, and their antiproliferative and antioxidant activities," Journal of Agricultural and Food Chemistry, vol. 54, no. 19, pp. 7069–7074, 2006.

63. D. Konopacka, K. Jesionkowska, D. Kruczyńska et al., "Apple and peach consumption habits across European countries," Appetite, vol. 55, no. 3, pp. 478–483, 2010.

64. A. G. Schauss, X. Wu, R. L. Prior et al., "Phytochemical and nutrient composition of the freeze-dried amazonian palm berry, Euterpe oleraceae Mart. (Acai)," Journal of Agricultural and Food Chemistry, vol. 54, no. 22, pp. 8598–8603, 2006.

65. A. G. Schauss, X. Wu, R. L. Prior et al., "Antioxidant capacity and other bioactivities of the freeze-dried Amazonian palm berry, Euterpe oleraceae Mart. (Acai)," Journal of Agricultural and Food Chemistry, vol. 54, no. 22, pp. 8604–8610, 2006.

66. O. Boyd, P. Weng, X. Sun et al., "Nectarine promotes longevity in Drosophila melanogaster," Free Radical Biology and Medicine, vol. 50, no. 11, pp. 1669–1678, 2011.

67. X. Sun, J. Seeberger, T. Alberico et al., "Açai palm fruit (Euterpe oleracea Mart.) pulp improves survival of flies on a high fat diet," Experimental Gerontology, vol. 45, no. 3, pp. 243–251, 2010.

68. P. Liedo, J. R. Carey, D. K. Ingram, and S. Zou, "The interplay among dietary fat, sugar, protein and acai (Euterpe oleracea Mart.) pulp in modulating lifespan and reproduction in a Tephritid fruit fly,"Experimental Gerontology, vol. 47, no. 7, pp. 536–539, 2012.

69. K. J. Pearson, J. A. Baur, K. N. Lewis et al., "Resveratrol delays age-related deterioration and mimics transcriptional aspects of dietary restriction without extending life span," Cell Metabolism, vol. 8, no. 2, pp. 157–168, 2008.

70. S. Zou, J. R. Carey, P. Liedo et al., "The prolongevity effect of resveratrol depends on dietary composition and calorie intake in a tephritid fruit fly," Experimental Gerontology, vol. 44, no. 6-7, pp. 472–476, 2009.

71. C. Wang, C. T. Wheeler, T. Alberico, et al., "The effect of resveratrol on lifespan depends on both gender and dietary nutrient composition in Drosophila melanogaster," Age. In press.

72. P. Liedo, J. R. Carey, D. K. Ingram, and S. Zou, "The interplay among dietary fat, sugar, protein and acai (Euterpe oleracea Mart.) pulp in modulating lifespan and reproduction in a Tephritid fruit fly,"Experimental Gerontology, vol. 47, no. 7, pp. 536–539, 2012.

73. N. G. Baydar, H. Baydar, and T. Debener, "Analysis of genetic relationships among Rosa damascenaplants grown in Turkey by using AFLP and microsatellite markers," Journal of Biotechnology, vol. 111, no. 3, pp. 263–267, 2004.

74. T. Hongratanaworakit, "Relaxing effect of rose oil on humans," Natural Product Communications, vol. 4, no. 2, pp. 291–296, 2009.

75. N. Kumar, P. Bhandari, B. Singh, A. P. Gupta, and V. K. Kaul, "Reversed phase-HPLC for rapid determination of polyphenols in

flowers of rose species," Journal of Separation Science, vol. 31, no. 2, pp. 262–267, 2008.

76. E. Basim and H. Basim, "Antibacterial activity of Rosa damascena essential oil," Fitoterapia, vol. 74, no. 4, pp. 394–396, 2003. · ·
77. R. Ramezani, A. Moghimi, H. Rakhshandeh, H. Ejtehadi, and M. Kheirabadi, "The effect of Rosa damascena essential oil on the amygdala electrical kindling seizures in rat," Pakistan Journal of Biological Sciences, vol. 11, no. 5, pp. 746–751, 2008.
78. N. Shokouhinejad, M. Emaneini, M. Aligholi, and F. Jabalameli, "Antimicrobial effect of Rosa damascena extract on selected endodontic pathogens," Journal of the California Dental Association, vol. 38, no. 2, pp. 123–126, 2010.
79. S. Awale, C. Tohda, Y. Tezuka, et al., "Protective effects of Rosa damascena and its active constituent on abeta(25–35)-induced neuritic atrophy," Evidence-based Complementary and Alternative Medicine, vol. 2011, Article ID 131042, 2011.
80. S. E. Schriner, N. S. Katoozi, K. Q. Pham, M. Gazarian, A. Zarban, and M. Jafari, "Extension ofDrosophila lifespan by Rosa damascena associated with an increased sensitivity to heat," Biogerontology, vol. 13, no. 2, pp. 105–117, 2012.
81. J. H. Weisburger, "Chemopreventive effects of cocoa polyphenols on chronic diseases," Experimental Biology and Medicine, vol. 226, no. 10, pp. 891–897, 2001.
82. P. Martorell, J. V. Forment, R. De Llanos et al., "Use of Saccharomyces cerevisiae and Caenorhabditis elegans as model organisms to study the effect of cocoa polyphenols in the resistance to oxidative stress," Journal of Agricultural and Food Chemistry, vol. 59, no. 5, pp. 2077–2085, 2011.
83. N. T. Zaveri, "Green tea and its polyphenolic catechins: medicinal uses in cancer and noncancer applications," Life Sciences, vol. 78, no. 18, pp. 2073–2080, 2006.
84. S. M. Sagar, D. Yance, and R. K. Wong, "Natural health products that inhibit angiogenesis: a potential source for investigational new agents to treat cancer-part 1," Current Oncology, vol. 13, no. 1, pp. 14–26, 2006.
85. J. A. Yang, J. H. Choi, and S. J. Rhee, "Effects of green tea catechin on phospholipase A2 activity and antithrombus in streptozotocin diabetic rats," Journal of Nutritional Science and Vitaminology,

vol. 45, no. 3, pp. 337–346, 1999.

86. P. G. Pietta, "Flavonoids as antioxidants," Journal of Natural Products, vol. 63, no. 7, pp. 1035–1042, 2000.

87. S. G. Khan, S. K. Katiyar, R. Agarwal, and H. Mukhtar, "Enhancement of antioxidant and phase II enzymes by oral feeding of green tea polyphenols in drinking water to SKH-1 hairless mice: possible role in cancer chemoprevention," Cancer Research, vol. 52, no. 14, pp. 4050–4052, 1992.

88. Y. M. Li, H. Y. E. Chan, Y. Huang, and Z. Y. Chen, "Green tea catechins upregulate superoxide dismutase and catalase in fruit flies," Molecular Nutrition and Food Research, vol. 51, no. 5, pp. 546–554, 2007.

89. Y. M. Li, H. Y. E. Chan, X. Q. Yao, Y. Huang, and Z. Y. Chen, "Green tea catechins and broccoli reduce fat-induced mortality in Drosophila melanogaster," Journal of Nutritional Biochemistry, vol. 19, no. 6, pp. 376–383, 2008.

90. S. Abbas and M. Wink, "Epigallocatechin gallate from green tea (Camellia sinensis) increases lifespan and stress resistance in Caenorhabditis elegans," Planta Medica, vol. 75, no. 3, pp. 216–221, 2009.

91. L. Zhang, G. Jie, J. Zhang, and B. Zhao, "Significant longevity-extending effects of EGCG onCaenorhabditis elegans under stress," Free Radical Biology and Medicine, vol. 46, no. 3, pp. 414–421, 2009.

92. K. Zarse, S. Jabin, and M. Ristow, "L: -Theanine extends lifespan of adult Caenorhabditis elegans,"European Journal of Nutrition, vol. 51, no. 6, pp. 765–768, 2012.

93. T. I. Kim, Y. K. Lee, S. G. Park et al., "l-Theanine, an amino acid in green tea, attenuates β-amyloid-induced cognitive dysfunction and neurotoxicity: reduction in oxidative damage and inactivation of ERK/p38 kinase and NF-κB pathways," Free Radical Biology and Medicine, vol. 47, no. 11, pp. 1601–1610, 2009.

94. J. Bryan, "Psychological effects of dietary components of tea: caffeine and L-theanine," Nutrition Reviews, vol. 66, no. 2, pp. 82–90, 2008.

95. Y. Sadzuka, T. Sugiyama, A. Miyagishima, Y. Nozawa, and S. Hirota, "The effects of theanine, as a novel biochemical

modulator, on the antitumor activity of adriamycin," Cancer Letters, vol. 105, no. 2, pp. 203–209, 1996.

96. H. Yokogoshi, Y. Kato, Y. M. Sagesaka, T. Takihara-Matsuura, T. Kakuda, and N. Takeuchi, "Reduction effect of theanine on blood pressure and brain 5-hydroxyindoles in spontaneously hypertensive rats," Bioscience, Biotechnology, and Biochemistry, vol. 59, no. 4, pp. 615–618, 1995.
97. G. Buckland, A. Agudo, N. Travier, et al., "Adherence to the Mediterranean diet reduces mortality in the Spanish cohort of the European Prospective Investigation into Cancer and Nutrition (EPIC-Spain)," British Journal of Nutrition, vol. 106, no. 10, pp. 1581–1591, 2011.
98. M. Gerber, "Biofactors in the Mediterranean diet," Clinical Chemistry and Laboratory Medicine, vol. 41, no. 8, pp. 999–1004, 2003.
99. P. Lagiou, D. Trichopoulos, S. Sandin et al., "Mediterranean dietary pattern and mortality among young women: a cohort study in Sweden," British Journal of Nutrition, vol. 96, no. 2, pp. 384–392, 2006.
100. J. C. Stoclet, T. Chataigneau, M. Ndiaye et al., "Vascular protection by dietary polyphenols," European Journal of Pharmacology, vol. 500, no. 1–3, pp. 299–313, 2004.
101. A. Trichopoulou, "Modified Mediterranean diet and survival: EPIC-elderly prospective cohort study,"British Medical Journal, vol. 330, no. 7498, pp. 991–995, 2005.
102. A. Bendini, L. Cerretani, A. Carrasco-Pancorbo et al., "Phenolic molecules in virgin olive oils: a survey of their sensory properties, health effects, antioxidant activity and analytical methods. An overview of the last decade," Molecules, vol. 12, no. 8, pp. 1679–1719, 2007.
103. A. Cañuelo, B. Gilbert-López, P. Pacheco-Liñán, E. Martínez-Lara, E. Siles, and A. Miranda-Vizuete, "Tyrosol, a main phenol present in extra virgin olive oil, increases lifespan and stress resistance inCaenorhabditis elegans," Mechanisms of Ageing and Development, vol. 133, no. 8, pp. 563–574, 2012.·
104. A. L. Hsu, C. T. Murphy, and C. Kenyon, "Regulation of aging and age-related disease by DAF-16 and heat-shock factor," Science, vol. 300, no. 5622, pp. 1142–1145, 2003.

105. A. W. Boots, G. R. M. M. Haenen, and A. Bast, "Health effects of quercetin: from antioxidant to nutraceutical," European Journal of Pharmacology, vol. 585, no. 2-3, pp. 325–337, 2008.

106. K. Pietsch, N. Saul, R. Menzel, S. R. Stürzenbaum, and C. E. W. Steinberg, "Quercetin mediated lifespan extension in Caenorhabditis elegans is modulated by age-1, daf-2, sek-1 and unc-43,"Biogerontology, vol. 10, no. 5, pp. 565–578, 2009.

107. N. Saul, K. Pietsch, R. Menzel, and C. E. W. Steinberg, "Quercetin-mediated longevity inCaenorhabditis elegans: is DAF-16 involved?" Mechanisms of Ageing and Development, vol. 129, no. 10, pp. 611–613, 2008.

108. N. Saul, K. Pietsch, R. Menzel, S. R. Stürzenbaum, and C. E. W. Steinberg, "The longevity effect of tannic acid in Caenorhabditis elegans: disposable soma meets hormesis," Journals of Gerontology, vol. 65, no. 6, pp. 626–635, 2010.

109. F. Surco-Laos, J. Cabello, E. Gómez-Orte, et al., "Effects of O-methylated metabolites of quercetin on oxidative stress, thermotolerance, lifespan and bioavailability on Caenorhabditis elegans," Food & Function, vol. 2, no. 8, pp. 445–456, 2011.

110. A. Kampkötter, C. Timpel, R. F. Zurawski, et al., "Increase of stress resistance and lifespan ofCaenorhabditis elegans by quercetin," Comparative Biochemistry and Physiology Part B, vol. 149, no. 2, pp. 314–323, 2008.

111. K. Yazawa, T. Kihara, H. Shen, Y. Shimmyo, T. Niidome, and H. Sugimoto, "Distinct mechanisms underlie distinct polyphenol-induced neuroprotection," FEBS Letters, vol. 580, no. 28-29, pp. 6623–6628, 2006.

112. T. J. Kim, J. L. Silva, and Y. S. Jung, "Antibacterial activity of fresh and processed red muscadine juice and the role of their polar compounds on Escherichia coli O157:H7," Journal of Applied Microbiology, vol. 107, no. 2, pp. 533–539, 2009.

113. T. Taguri, T. Tanaka, and I. Kouno, "Antimicrobial activity of 10 different plant polyphenols against bacteria causing food-borne disease," Biological and Pharmaceutical Bulletin, vol. 27, no. 12, pp. 1965–1969, 2004.

114. M. T. Huang, R. L. Chang, and A. W. Wood, "Inhibition of the mutagenicity of bay-region diol-epoxides of polycyclic aromatic hydrocarbons by tannic acid, hydroxylated anthraquinones and

hydroxylated cinnamic acid derivatives," Carcinogenesis, vol. 6, no. 2, pp. 237–242, 1985.

115. S. Taffetani, Y. Ueno, F. Meng et al., "Tannic acid inhibits cholangiocyte proliferation after bile duct ligation via a cyclic adenosine 5′,3′-monophosphate-dependent pathway," American Journal of Pathology, vol. 166, no. 6, pp. 1671–1679, 2005.

116. R. G. Andrade Jr., J. S. Ginani, G. K. B. Lopes, F. Dutra, A. Alonso, and M. Hermes-Lima, "Tannic acid inhibits invitro iron-dependent free radical formation," Biochimie, vol. 88, no. 9, pp. 1287–1296, 2006.·

117. G. K. B. Lopes, H. M. Schulman, and M. Hermes-Lima, "Polyphenol tannic acid inhibits hydroxyl radical formation from Fenton reaction by complexing ferrous ions," Biochimica et Biophysica Acta, vol. 1472, no. 1-2, pp. 142–152, 1999.

118. K. Pietsch, N. Saul, S. C. Swain, R. Menzel, C. E. Steinberg, and S. R. Stürzenbaum, "Meta-analysis of global transcriptomics suggests that conserved genetic pathways are responsible for quercetin and tannic acid mediated longevity in C. elegans," Frontiers in Genetics, vol. 3, article 48, 2012.

119. J. L. Johannes andV. L. S. Cilliers, "Characterization of the products of nonenzymic autoxidative phenolic reactions in a caffeic acid model system," Journal of Agricultural and Food Chemistry, vol. 39, no. 7, pp. 1298–1303, 1991.

120. M. R. Al-Sereiti, K. M. Abu-Amer, and P. Sen, "Pharmacology of rosemary (Rosmarinus officinalisLinn.) and its therapeutic potentials," Indian Journal of Experimental Biology, vol. 37, no. 2, pp. 124–130, 1999.

121. A. Sahu, N. Rawal, and M. K. Pangburn, "Inhibition of complement by covalent attachment of rosmarinic acid to activated C3b," Biochemical Pharmacology, vol. 57, no. 12, pp. 1439–1446, 1999.·

122. M. A. Kelm, M. G. Nair, G. M. Strasburg, and D. L. DeWitt, "Antioxidant and cyclooxygenase inhibitory phenolic compounds from Ocimum sanctum Linn," Phytomedicine, vol. 7, no. 1, pp. 7–13, 2000.

123. W. Zheng and S. Y. Wang, "Antioxidant activity and phenolic compounds in selected herbs," Journal of Agricultural and Food

Chemistry, vol. 49, no. 11, pp. 5165–5170, 2001.

124. K. Pietsch, N. Saul, S. Chakrabarti, S. R. Stürzenbaum, R. Menzel, and C. E. W. Steinberg, "Hormetins, antioxidants and prooxidants: defining quercetin-, caffeic acid- and rosmarinic acid-mediated life extension in C. elegans," Biogerontology, vol. 12, no. 4, pp. 329–347, 2011.

125. T. Eisenberg, H. Knauer, A. Schauer, et al., "Induction of autophagy by spermidine promotes longevity," Nature Cell Biology, vol. 11, no. 11, pp. 1305–1314, 2009.

126. E. Morselli, G. Mariño, M. V. Bennetzen et al., "Spermidine and resveratrol induce autophagy by distinct pathways converging on the acetylproteome," Journal of Cell Biology, vol. 192, no. 4, pp. 615–629, 2011.

127. B. Levine and G. Kroemer, "Autophagy in the pathogenesis of disease," Cell, vol. 132, no. 1, pp. 27–42, 2008.

128. G. Scalabrino and M. E. Ferioli, "Polyamines in mammalian aging: an oncological problem, too? A review," Mechanisms of Ageing and Development, vol. 26, no. 2-3, pp. 149–164, 1984.

129. M. Braunstein, R. E. Sobel, C. D. Allis, B. M. Turner, and J. R. Broach, "Efficient transcriptional silencing in Saccharomyces cerevisiae requires a heterochromatin histone acetylation pattern,"Molecular and Cellular Biology, vol. 16, no. 8, pp. 4349–4356, 1996.

130. E. Ford, R. Voit, G. Liszt, C. Magin, I. Grummt, and L. Guarente, "Mammalian Sir2 homolog SIRT7 is an activator of RNA polymerase I transcription," Genes and Development, vol. 20, no. 9, pp. 1075–1080, 2006.

131. R. Mostoslavsky, K. F. Chua, D. B. Lombard et al., "Genomic instability and aging-like phenotype in the absence of mammalian SIRT6," Cell, vol. 124, no. 2, pp. 315–329, 2006.

132. B. B. Aggarwal and K. B. Harikumar, "Potential therapeutic effects of curcumin, the anti-inflammatory agent, against neurodegenerative, cardiovascular, pulmonary, metabolic, autoimmune and neoplastic diseases," International Journal of Biochemistry and Cell Biology, vol. 41, no. 1, pp. 40–59, 2009.

133. B. B. Aggarwal and B. Sung, "Pharmacological basis for the role of curcumin in chronic diseases: an age-old spice with modern

targets," Trends in Pharmacological Sciences, vol. 30, no. 2, pp. 85–94, 2009.·

134. S. Bengmark, "Curcumin, an atoxic antioxidant and natural NFκB, cyclooxygenase-2, lipooxygenase, and inducible nitric oxide synthase inhibitor: a shield against acute and chronic diseases," Journal of Parenteral and Enteral Nutrition, vol. 30, no. 1, pp. 45–51, 2006.
135. R. K. Maheshwari, A. K. Singh, J. Gaddipati, and R. C. Srimal, "Multiple biological activities of curcumin: a short review," Life Sciences, vol. 78, no. 18, pp. 2081–2087, 2006.
136. H. Nishino, H. Tokuda, Y. Satomi et al., "Cancer prevention by antioxidants," BioFactors, vol. 22, no. 1–4, pp. 57–61, 2004.
137. K. S. Lee, B. S. Lee, S. Semnani et al., "Curcumin extends life span, improves health span, and modulates the expression of age-associated aging genes in Drosophila melanogaster," Rejuvenation Research, vol. 13, no. 5, pp. 561–570, 2010.
138. V. H. Liao, C. W. Yu, Y. J. Chu, W. H. Li, Y. C. Hsieh, and T. T. Wang, "Curcumin-mediated lifespan extension in Caenorhabditis elegans," Mechanisms of Ageing and Development, vol. 132, no. 10, pp. 480–487, 2011.
139. R. I. Morimoto, "Proteotoxic stress and inducible chaperone networks in neurodegenerative disease and aging," Genes and Development, vol. 22, no. 11, pp. 1427–1438, 2008.
140. R. R. Kopito and D. Ron, "Conformational disease," Nature Cell Biology, vol. 2, no. 11, pp. E207–E209, 2000.
141. D. Eisenberg and M. Jucker, "The amyloid state of proteins in human diseases," Cell, vol. 148, no. 6, pp. 1188–1203, 2012.
142. S. Alavez, M. C. Vantipalli, D. J. S. Zucker, I. M. Klang, and G. J. Lithgow, "Amyloid-binding compounds maintain protein homeostasis during ageing and extend lifespan," Nature, vol. 472, no. 7342, pp. 226–230, 2011.
143. D. A. Skorupa, A. Dervisefendic, J. Zwiener, and S. D. Pletcher, "Dietary composition specifies consumption, obesity, and lifespan in Drosophila melanogaster," Aging Cell, vol. 7, no. 4, pp. 478–490, 2008.
144. S. Zou, P. Liedo, L. Altamirano-Robles et al., "Recording lifetime behavior and movement in an invertebrate model," PLoS ONE, vol. 6, no. 4, Article ID e18151, 2011.

145. J. R. Carey, L. G. Harshman, P. Liedo, H. G. Müller, J. L. Wang, and Z. Zhang, "Longevity-fertility trade-offs in the tephritid fruit fly, Anastrepha ludens, across dietary-restriction gradients," Aging Cell, vol. 7, no. 4, pp. 470–477, 2008.
146. G. D. Lee, M. A. Wilson, M. Zhu et al., "Dietary deprivation extends lifespan in Caenorhabditis elegans," Aging Cell, vol. 5, no. 6, pp. 515–524, 2006.
147. M. D. W. Piper, L. Partridge, D. Raubenheimer, and S. J. Simpson, "Dietary restriction and aging: a unifying perspective," Cell Metabolism, vol. 14, no. 2, pp. 154–160, 2011.
148. K. J. Vinknes, S. De Vogel, A. K. Elshorbagy et al., "Dietary intake of protein is positively associated with percent body fat in middle-aged and older adults," Journal of Nutrition, vol. 141, no. 3, pp. 440–446, 2011.
149. M. Meydani and S. T. Hasan, "Dietary polyphenols and obesity," Nutrients, vol. 2, no. 7, pp. 737–751, 2010.

Chapter 5

Antioxidant Activity, Total Phenolic and Total Flavonoid Contents of Whole Plant Extracts Torilis leptophylla L

Naima Saeed, Muhammad R Khan, and Maria Shabbir

Department of Biochemistry, Faculty of Biological Sciences, Quaid-i-Azam University, Islamabad, 45320, Pakistan

ABSTRACT

Background

The aim of this study was to screen various solvent extracts of whole plant of *Torilis leptophylla* to display potent antioxidant activity *in vitro* and *in vivo,* total phenolic and flavonoid contents in order to find possible sources for future novel antioxidants in food and pharmaceutical formulations.

Material and Methods

A detailed study was performed on the antioxidant activity of the methanol extract of whole plant of *Torilis leptophylla* (TLM) and its derived fractions {*n*-hexane (TLH), chloroform (TLC) ethyl acetate (TLE) *n*-butanol (TLB) and residual aqueous fraction (TLA)} by *in vitro* chemical analyses and carbon tetrachloride (CCl_4) induced hepatic injuries (lipid peroxidation and glutathione contents) in male Sprague-Dawley rat. The total yield, total phenolic (TPC) and total flavonoid contents (TFC) of all the fractions were also determined. TLM was also subjected to preliminary phytochemical screening test for various constituents.

Results

The total phenolic contents (TPC) (121.9±3.1 mg GAE/g extract) of TLM while total flavonoid contents (TFC) of TLE (60.9 ±2.2 mg RTE/g extract) were found significantly higher as compared to other solvent fractions. Phytochemical screening of TLM revealed the presence of alkaloids, anthraquinones, cardiac glycosides, coumarins, flavonoids, saponins, phlobatannins, tannins and terpenoids. The EC_{50} values based on the DPPH (41.0±1 μg/ml), ABTS (10.0±0.9 μg/ml) and phosphomolybdate (10.7±2 μg/ml) for TLB, hydroxyl radicals (8.0±1 μg/ml) for TLC, superoxide radicals (57.0±0.3 μg/ml) for TLM and hydrogen peroxide radicals (68.0±2 μg/ml) for TLE were generally lower showing potential antioxidant properties. A significant but marginal positive correlation was found between TPC and EC_{50} values for DPPH, hydroxyl, phosphomolybdate and ABTS, whereas another weak and positive correlation was determined between TFC and EC_{50}values for superoxide anion and hydroxyl radicals. Results of *in vivo* experiment revealed that administration of CCl_4 caused a significant increase in lipid peroxidation (TBARS) while decrease in GSH contents of liver. In contrast, TLM (200 mg/kg bw) and silymarin (50 mg/kg bw) co-treatment effectively prevented these alterations and maintained the antioxidant status.

Conclusion

Data from present results revealed that *Torilis leptophylla* act as an antioxidant agent due to its free radical scavenging and cytoprotective activity.

BACKGROUND

Since very old times, herbal medications have been used for relief of symptoms of disease [1]. Despite the great advances observed in modern medicine in recent decades, plants still make an important contribution to health care. Much interest, in medicinal plants however, emanates from their long use in folk medicines as well as their prophylactic properties, especially in developing countries. Large number of medicinal plants has been investigated for their antioxidant properties. Natural antioxidants either in the form of raw extracts or their chemical constituents are very effective to prevent the destructive processes caused by oxidative stress [2]. Although the toxicity profile of most medicinal plants have not been thoroughly evaluated, it is generally accepted that medicines derived from plant products are safer than their synthetic counterparts [3, 4].

Substantial evidence has accumulated and indicated key roles for reactive oxygen species (ROS) and other oxidants in causing numerous disorders and diseases. The evidence has brought the attention of scientists to an appreciation of antioxidants for prevention and treatment of diseases, and maintenance of human health [5]. Human body has an inherent antioxidative mechanism and many of the biological functions such as the anti-mutagenic, anti-carcinogenic, and anti-aging responses originate from this property [6,7]. Antioxidants stabilize or deactivate free radicals, often before they attack targets in biological cells [8]. Recently interest in naturally occurring antioxidants has considerably increased for use in food, cosmetic and pharmaceutical products, because they possess multifacetedness in their multitude and magnitude of activity and provide enormous scope in correcting imbalance [9,10].

The role of free radical reactions in disease pathology is well established and is known to be involved in many acute and chronic disorders in human beings, such as diabetes, atherosclerosis, aging, immunosuppression and neurodegeneration [11]. An imbalance between ROS and the inherent antioxidant capacity of the body, directed the use of dietary and /or medicinal supplements particularly during the disease attack. Studies on herbal plants, vegetables, and fruits have indicated the presence of antioxidants such as phenolics, flavonoids, tannins, and proanthocyanidins. The antioxidant contents of medicinal plants may contribute to the protection they offer from disease. The ingestion of natural antioxidants has been inversely associated with morbidity and mortality from degenerative disorders [6]. Liver diseases remain a serious health problem. It is well known that free radicals cause cell damage through mechanisms of covalent binding and lipid peroxidation with subsequent tissue injury. Antioxidant agents of natural origin have attracted special interest because of their free radical scavenging abilities [12]. The use of medicinal plants with high level of antioxidant constituents has been proposed as an effective therapeutic approach for hepatic damages [13].

Torilis leptophylla belonging to the Apiaceae family [14] or more commonly known as Bristle fruit Hedge parsley, is a forb of the genus *Torilis*, distributed in Asia, Europe and North Africa. It is represented in Iran by nine species [15]. *T. leptophylla* has been used in folk medicine for the treatment of gastrointestinal illnesses in Iran and Pakistan. The plant is highly effective against some pathogens thus confirming its use as disinfectant or antiseptic [16].The search for novel natural antioxidants of plant origin has ever since increased. It is not known which constituents of plant are associated in reducing the risk of chronic diseases, but antioxidants appear to play a major role in the protective effect of plant medicine. The present study was designed to investigate the TPC and TFC and to evaluate the antioxidant activities (*in vitro* and *in vivo*) of the various fractions of methanol extract of whole plant of *T. leptophylla*. The 95% methanol extract was also subjected to phytochemical screening to determine the presence of alkaloids, anthraquinones, cardiac glycosides, coumarins, flavonoids, saponins, phlobatannins, tannins and terpenoids.

METHODS

Plant Collection

The plant was collected in April 2011 from Islamabad, Pakistan. The plant material was botanically identified by Dr. Mir Ajab Khan, Department of Plant Sciences, Quaid-i-Azam University Islamabad. A voucher specimen (accession #023541) was deposited at the Herbarium of Pakistan Museum of Natural History Islamabad.

Extract Preparation

The fresh, whole plant (3 kg) was collected and shade dried to obtain 500 g dry sample which was later coarsely powdered in a Willy Mill to 60-mesh size and used for solvent extraction. For sample preparation, 500 g of dried sample were extracted twice (2000 ml for each) with 95% methanol at 25°C for 48 h and concentrated using a rotary evaporator (Panchun Scientific Co., Kaohsiung, Taiwan) under reduced pressure at 40°C to yield the TLM (11.5%). The residue was suspended in water (50 ml) and partitioned successively with *n*-hexane, chloroform, ethyl acetate, *n*-butanol (a total of two aliquots of 100 ml each) and soluble residual aqueous fraction yielding respectively the TLH (5.4%), TLC (4.3%), TLE (6.1%), TLB (4.8%) and TLA (8.2%) [17].

Chemicals

Ascorbic acid; aluminum chloride, 2, 2′ - azino-bis-(3-ethylbenzothiazoline-6-sulphonic acid) (ABTS); ferric chloride ($FeCl_3$); Folin-Ciocalteu; bovine serum albumin (BSA); potassium persulphate; 2,2′-diphenyl-1-picrylhydrazyl (DPPH); nitro blue tetrazolium (NBT); phenazine methosulphate (PMS); reduced glutathione (GSH); 1,2-dithio-bis nitro benzoic acid (DTNB); sulphosalicylic acid; thiobarbituric acid (TBA) and trichloroacetic acid (TCA) were purchased from Sigma Co. (St. Louis, MO, USA). Sulphuric acid; 2-deoxyribose; riboflavin; sodium carbonate (Na_2CO_3); sodium hydroxide (NaOH); sodium nitrite ($NaNO_2$); disodium hydrogen phosphate (Na_2HPO_4) and hydrogen peroxide (H_2O_2) were obtained from Wako Co. (Osaka,

Japan). Potassium ferricyanide [$K_3Fe(CN)_6$]; triflouroacetic acid; sodium dihydrogen phosphate (NaH_2PO_4) and all solvents *n*-hexane (99.8%); chloroform (99.8%); ethyl acetate (99.8%) and *n*-butanol (99.8%) used were of analytical grade and purchased from Merck Co. (Darmstadt, Germany). Distilled deionized water (dd. H_2O) was prepared by Ultrapure TM water purification system (Lotun Co., Ltd., Taipei, Taiwan).

In Vitro Studies

Antioxidant Assays

Each sample was dissolved in 95% methanol to make a concentration of 1 mg/ml and then diluted to prepare the series concentrations for antioxidant assays. Reference chemicals were used for comparison in all assays.

DPPH Radical Scavenging Activity Assay

The free radical scavenging activity of the fractions was measured *in vitro* by 2,2'- diphenyl-1-picrylhydrazyl (DPPH) assay according to the method described earlier [18,19]. The stock solution was prepared by dissolving 24 mg DPPH with 100 ml methanol and stored at 20°C until required. The working solution was obtained by diluting DPPH solution with methanol to attain an absorbance of about 0.98±0.02 at 517 nm using the spectrophotometer. A 3 ml aliquot of this solution was mixed with 100 μl of the sample at various concentrations (10 - 500 μg/ml). The reaction mixture was shaken well and incubated in the dark for 15 min at room temperature. Then the absorbance was taken at 517 nm. The control was prepared as above without any sample. The scavenging activity was estimated based on the percentage of DPPH radical scavenged as the following equation:

$$\text{Scavening effect } (\%) = \left[(\text{controlabsorance} - \text{sampleabsorance}) / (\text{contraolasbsorbance})\right] \times 100 \quad (1)$$

Superoxide Anion Scavenging Assay

The assay for superoxide anion radical scavenging activity was supported by riboflavin-light-NBT system [20]. Briefly, 1 ml of sample was taken at different concentrations (25 to 500 μg/ml) and mixed with 0.5 ml of phosphate buffer (50 mM, pH 7.6), 0.3 ml riboflavin (50 mM), 0.25 ml PMS (20 mM), and 0.1 ml NBT (0.5 mM). Reaction was started by illuminating the reaction mixture using a fluorescent lamp. After 20 min of incubation, the absorbance was measured at 560 nm. Ascorbic acid was used as standard. The scavenging ability of the plant extract was determined by the following equation:

$$\text{Scavening effect (\%)} = \left(1 - \text{absorbance of sample} / \text{absorbance of control}\right) \times 100 \quad (2)$$

Phosphomolybdate Assay (Total Antioxidant Capacity)

The total antioxidant capacity of the fractions was determined by phosphomolybdate method using ascorbic acid as a standard [21]. An aliquot of 0.1 ml of sample solution was mixed with 1 ml of reagent solution (0.6 M sulphuric acid, 28 mM sodium phosphate and 4 mM ammonium molybdate). The tubes were capped and incubated in a water bath at 95°C for 90 min. After the samples had cooled to room temperature, the absorbance of the mixture was measured at 765 nm against a blank. A typical blank contained 1 ml of the reagent solution and the appropriate volume of the solvent and incubated under the same conditions. Ascorbic acid was used as standard. The antioxidant capacity was estimated using following formula:

$$\text{Antioxidant effect (\%)} = \left[(\text{control absorbance} - \text{sample absorbance}) / (\text{control absorbance})\right] \times 100 \quad (3)$$

Hydroxyl Radical Scavenging Assay

Hydroxyl radical scavenging activity was measured by the ability of the different fractions of *T.leptophylla* extract to scavenge the hydroxyl radicals generated by the Fe^{3+}-ascorbate-EDTA-H_2O_2 system (Fenton reaction) [22]. The reaction mixture contained; 500 µl of 2-deoxyribose (2.8 mM) in phosphate buffer (50 mM, pH 7.4), 200 µl of premixed ferric chloride (100 mM) and EDTA (100 mM) solution (1:1; v/v), 100 µl of H_2O_2 (200 mM) with or without the extract solution (100 µl). The reaction was triggered by adding 100 µl of 300 mM ascorbate and incubated for 1 h at 37°C. 0.5 ml of the reaction mixture was added to 1 ml of TCA (2.8%; w/v; aqueous solution), then 1 ml of 1% aqueous TBA were added to the reaction mixture. The mixture was heated for 15 min on a boiling water bath. After the mixture being cooled the absorbance at 532 nm was noted against a blank (the same solution but without reagent). The scavenging activity on hydroxyl radical was calculated as follows:

$$\text{Scavenging activity}(\%) = \left(1 - \text{absorbance of sample} / \text{absorbance of control}\right) \times 100 \quad (4)$$

Hydrogen Peroxide Scavenging Activity

Hydrogen peroxide solution (2 mM) was prepared in 50 mM phosphate buffer (pH 7.4). Aliquots (0.1 ml) of different fractions was transferred into the test tubes and their volumes were made up to 0.4 ml with 50 mM phosphate buffer (pH 7.4) After addition of 0.6 ml hydrogen peroxide solution, tubes were vortexed and absorbance of the hydrogen peroxide at 230 nm was determined after 10 min, against a blank [23]. The abilities to scavenge the hydrogen peroxide were calculated using the following equation:

$$\text{Hydrogen peroxide scavenging activity} = \left(1 - \text{absorbance of sample} / \text{absorbance of sample}\right) \times 100 \quad (5)$$

ABTS Radical Scavenging Activity

The 2,2′-azinobis (3-ethylbenzthiazoline-6-sulphonic acid), commonly called ABTS cation scavenging activity was performed [24]. Briefly, ABTS solution (7 mM) was reacted with potassium persulfate (2.45 mM) solution and kept for overnight in the dark to yield a dark coloured solution containing ABTS radical cations. Prior to use in the assay, the ABTS radical cation was diluted with 50% methanol for an initial absorbance of about 0.70±0.02 at 745 nm, with temperature control set at 30°C. Free radical scavenging activity was assessed by mixing 300 μl of test sample with 3.0 ml of ABTS working standard in a microcuvette. The decrease in absorbance was measured exactly one minute after mixing the solution, then up to 6 min. The percentage inhibition was calculated according to the formula:

$$\text{Scavenging effect}(\%) = \left[(\text{control absorbance} - \text{sample absorbance}) / (\text{control absorbance})\right] \times 100 \quad (6)$$

The antioxidant capacity of test samples was expressed as EC_{50} (anti-radical activity), the concentration necessary for 50% reduction of ABTS [25].

Reducing Power

The reducing power was based on Fe (III) to Fe (II) transformation in the presence of the solvent fractions [26]. The Fe (II) can be monitored by measuring the formation of Perl's Prussian blue at 700 nm. Various concentrations of the sample (2 ml) were mixed with 2 ml of phosphate buffer (0.2 M, pH 6.6) and 2 ml of potassium ferricyanide (10 mg/ml). The mixture was incubated at 50°C for 20 min followed by addition of 2 ml of trichloroacetic acid (100 mg/l). The mixture was centrifuged at 3000 rpm for 10 min to collect the upper layer of the solution. A volume of 2 ml from each of the mixture earlier mentioned was mixed with 2 ml of distilled water and 0.4 ml of 0.1% (w/v) fresh ferric chloride. After 10 min reaction, the absorbance was measured at 700 nm. Higher absorbance of the reaction mixture indicates a higher reducing power.

Estimation of Total Phenolic Content

The total phenolic content was determined by the spectrophotometric method [27]. In brief, a 1 ml of sample (1 mg/ml) was mixed with 1 ml of Folin-Ciocalteu's phenol reagent. After 5 min, 10 ml of a 7% Na_2CO_3 solution was added to the mixture followed by the addition of 13 ml of deionized distilled water and mixed thoroughly. The mixture was kept in the dark for 90 min at 23°C, after which the absorbance was read at 750 nm. The TPC was determined from extrapolation of calibration curve which was made by preparing gallic acid solution. The estimation of the phenolic compounds was carried out in triplicate. The TPC was expressed as milligrams of gallic acid equivalents (GAE) per g of dried sample.

Estimation of Total Flavonoid Content

Total flavonoid content was determined following a method by Park *et al* (2008) [28]. In a 10 ml test tube, 0.3 ml of extracts, 3.4 ml of 30% methanol, 0.15 ml of $NaNO_2$ (0.5 M) and 0.15 ml of $AlCl_3.6H_2O$ (0.3 M) were mixed. After 5 min, 1 ml of NaOH (1 M) was added. The solution was mixed well and the absorbance was measured against the reagent blank at 506 nm. The standard curve for total flavonoids was made using rutin standard solution (0 to 100 mg/l) under the same procedure as earlier described. The total flavonoids were expressed as milligrams of rutin equivalents per g of dried fraction.

Phytochemical Screening of TLM

Phytochemical screening of TLM for the presence of alkaloids, anthraquinones, cardiac glycosides, coumarins, flavonoids, saponins, phlobatannins, tannins and terpenoids was carried out.

Test for Alkaloids

0.4 g of TLM was stirred with 8 ml of 1% HCl and the mixture was warmed and filtered [29]. 2 ml of filtrate was treated separately with (a) with few drops of potassium mercuric iodide (Mayer's reagent) and (b) potassium bismuth (Dragendroff's reagent). Turbidity or precipitation with either of these reagents was taken as evidence for existence of alkaloids.

Test for Saponins

The ability of saponins to produce emulsion with oil was used for the screening test [29]. 20 mg of TLM was boiled in 20 ml of distilled water in a water bath for five min and filtered. 10 ml of the filtrate was mixed with 5 ml of distilled water and shaken vigorously for froth formation. 3 drops of olive oil were mixed with froth, shaken vigorously and observed for emulsion development.

Test for Terpenoids

Presence of terpenoids in TLM was carried out by taking 5 ml (1 mg/ml) of TLM and mixed with 2 ml of chloroform, followed by 3 ml of concentrated H_2SO_4. A reddish brown colouration of the interface confirmed the presence of terpenoids [29].

Test for Anthraquinones

200 mg of TLM was boiled with 6 ml of 1% HCl and filtered. The filtrate was shaken with 5 ml of benzene, filtered and 2 ml of 10% ammonia solution was added to the filtrate. The mixture was shaken and the presence of a pink, violet or red colour in the ammoniacal phase indicated the presence of free hydroxyl anthraquinones [30].

Cardiac Glycosides Determination

5 ml (10 mg/ml in methanol) of TLM was mixed with 2 ml of glacial acetic acid having one drop of $FeCl_3$ solution. To the mixture obtained 1 ml of concentrated H_2SO_4 was added to form a layer. The presence

of brown ring of the interface indicated deoxy sugar characteristic of cardiac glycosides [30].

Test for Coumarins

In a small test tube, 300 mg of TLM was covered with filter paper moistened with 1 N NaOH. The test tube was placed for few minutes in a boiling water bath. After removing the filter paper it was examined under UV light, yellow florescence indicated the presence of coumarins [30].

Test for Phlobatannins

80 mg of TLM was boiled in 1% aqueous hydrochloric acid; the deposition of a red precipitate indicated the presence of phlobatannins [30].

Test for Flavonoids

50 mg of TLM was suspended in 100 ml of distilled water to get the filtrate. 5 ml of dilute ammonia solution was added to 10 ml of filtrate followed by few drops of concentrated H_2SO_4. Presence of flavonoids was confirmed by yellow colouration [31].

Test for Tannins

50 mg of TLM was boiled in 20 ml of distilled water and filtered. A few drops of 0.1% $FeCl_3$ was added in filtrate and observed for colour change; brownish green or a blue-black colouration was taken as evidence for the presence of tannins [31].

In Vivo Studies

Animals and Treatment

Studies were carried out using male Sprague Dawley rats weighing 180±10 g. The animals were grouped and housed in polyacrylic cages with not more than six animals per cage and maintained under standard laboratory conditions. They had free access to standard diet and fresh water *ad libitum*. For acute toxicity studies, 30 rats were divided randomly into 5 groups, each comprising 6 animals. Group (I) the controls received only vehicles; olive oil (0.5 ml/kg bw) and DMSO (0.5 ml/kg bw) and fed with a normal diet for 7 days. Group II (induction controls) received single dose of CCl_4 (CCl_4 + Olive oil in 1:1 ratio; 2 ml/kg bw; i.p) on day 1 and day 7 of the experiment. Group III and IV received TLM (200 mg/kg bw; i.p) and silymarin (25 mg/kg, i.p) as the standard reference drug, once in a day for 7 days respectively along with the intraperitoneal administration of CCl_4 on day 1 and 7. Group V received only TLM (200 mg/kg; ip) once in a day for 7 days. At the end of 7 days, 24 h of the last treatment, all the animals were anesthetized in an ether chamber. The liver was removed and placed at 4°C after perfusion with ice cold saline. Intraperitoneal route of administration is selected because it is more rapid and predictable absorption than oral administration. This route is also advantageous because the drug is not inactivated or destroyed as may happen in the gastrointestinal tract and usually smaller doses are required because large concentrations of drug is achieved at the affected site. All experimental procedures involving animals were conducted in accordance with the guidelines of National Institutes of Health (NIH guidelines Islamabad, Pakistan). The study protocol was approved (No. 0241) by Ethical Committee of Quaid-i-Azam University, Islamabad.

Determination of in Vivo Antioxidant Activity

10% homogenate of liver tissue was prepared in 100 mM KH_2PO_4 buffer containing 1 mM EDTA (pH 7.4) and centrifuged at 12,000 × g for 30 min at 4°C. The supernatant was collected and used for the following experiments as described below.

Determination of Thiobarbituric Acid Reactive Substances (TBARS)

Malondialdehyde in liver homogenate was determined by reaction with thiobarbituric acid (TBA). Briefly; 1.0 ml reaction assay was consisted of 0.58 ml phosphate buffer (0.1 M; pH 7.4), 0.2 ml liver supernatant and following addition of 1.0 ml 0.67% thiobarbituric acid, all the tubes were placed in boiling water bath for 20 min and then shifted to crushed ice-bath before centrifuging at 2500 × g for 10 min. The amount of TBARS formed in each of the samples was assessed by measuring optical density of the supernatant at 535 nm using spectrophotometer against a reagent blank. The results were expressed as nM TBARS/min/mg tissue at 37°C using molar extinction coefficient of 1.56×10^5 $M^{-1}cm^{-1}$ [32].

Determination of Reduced Glutathione (GSH)

Reduced glutathione in liver homogenate was determined by reaction with 1,2-dithio-bis nitro benzoic acid (DTNB). Briefly, 1.0 ml of supernatant was precipitated with 1.0 ml of (4%) sulphosalicylic acid. The samples were kept at 4°C for 1 h and then centrifuged at 1200 × g for 20 min at 4°C. The total volume of 3.0 ml assay mixture contained 0.1 ml filtered aliquot, 2.7 ml phosphate buffer (0.1 M; pH 7.4) and 0.2 ml of 1,2-dithio-bis nitro benzoic acid (DTNB, 100 mM). The yellow colour developed was read immediately at 412 nm on a Smart- SpecTM plus Spectrophotometer. It was expressed as µM GSH/g tissue [33].

Statistical Analysis

Data are expressed as mean ± SD from three separate observations. For in vitro antioxidant assays one way ANOVA test followed by Tukey's test ($P < 0.05$) was used to analyze the differences among EC_{50} of various fractions for different antioxidant assays. The EC_{50} values were determined using the Graph Pad Prism 5 software. Data on biochemical investigations of *in vivo* experiments were analyzed by one-way (ANOVA) and the group means were compared by Dunnet's Multiple Range Test. A probability of $P < 0.05$ was considered as significant.

RESULTS

Phytochemical Screening

Phytochemical screening of TLM demonstrated the presence of alkaloids, anthraquinones, cardiac glycosides, coumarins, flavonoids, saponins, phlobatannins, tannins and terpenoids.

In Vitro Antioxidant Activity

Antioxidant capacity of TLM and its derived fractions was examined using seven different assays.

DPPH Radical Scavenging Activity

Figure 1A shows that the scavenging effects of samples on DPPH radical and were in the following order: TLB > TLE > TLC > TLM >TLA > TLH. The EC_{50} values of scavenging DPPH radicals for the TLB and TLE were 41.0±1 and 62.0±2 µg/ml, respectively (Table 1). Though the antioxidant potential of fractions was found to be low ($P < 0.05$) than those of ascorbic acid and rutin, the study revealed that TLB and TLE have prominent antioxidant activity; the presence of phenolic compounds (containing phenolic hydroxyls) are mainly found in these two fractions and could be attributable to the observed high antiradical properties of these fractions.

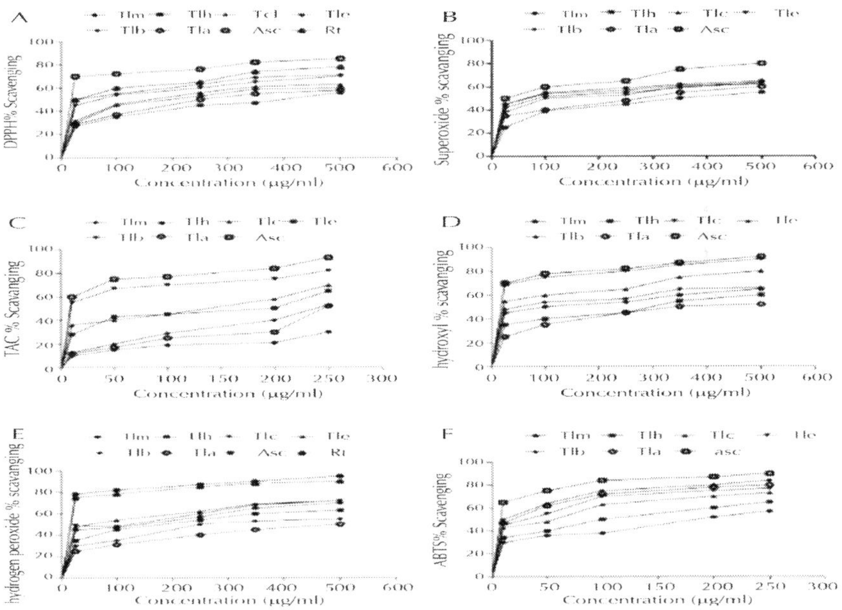

Figure 1: Antioxidant activities of different extracts from the methanol extract of whole plant of *T. leptophylla* by different solvents at various concentrations.

Each value represents a mean ± SD (n = 3): (**A**) DPPH radical scavenging activity, (**B**) Superoxide radical scavenging activity, (**C**) Total antioxidant capacity, (**D**) Hydroxyl radical scavenging activity, (**E**) Hydrogen peroxide radical scavenging activity, (F) ABTS radical scavenging activity. Tlm., methanol fraction, Tlh., *n*-hexane fraction, Tlc., chloroform fraction, Tle., ethyl acetate fraction, Tlb., *n*-butanol fraction, Tla., residual aqueous fraction, Asc., ascorbic acid, Rt., rutin.

Table 1: Radical scavenging activities of *T. leptophylla* fractions at different concentrations

EC50values(μg/ml) of radical scavenging						
Plant extracts/ chemical	DPPH radical	Superoxide radical	Phos phomoly bdate Assay	Hydroxyl radical	Hydrogen peroxide	ABTS radical
Methanol extract	189± 4e	55.0±0.3d	236±1e	76.0±2d	130±5e	179±3e
n-Hexane fraction	395± 5f	370±4f	247±3e	281±2e	200±2e	50.0±1c
Chloroform fraction	162± 3e	68.0±2d	135±4e	8.0±1a	122±5e	57.0±2d
Ethyl acetate fraction	62.0± 2d	145±1e	196±4e	27.0±1b	68.0±2d	34.0±3b
n-Butanol fraction	41.0± 1c	92.0±1d	10.7±2a	62.0±2d	254±3e	10.0±0.9a
Residual aqueous fraction	264± 4e	269±3e	237±3e	345±4f	431±5f	13.0±0.2a
Ascorbic acid	10.0± 3a	34.0±3b	8.0±0.7a	6.0±2a	9.0±0.3a	8.0±1a
Rutin	29.0± 1b	-	-	-	10.0±0.3a	-

Each value in the table is represented as mean ± SD (n = 3). Values in the same column followed by a different letter (a-f) are significantly different ($p < 0.05$). -, not determined.

Saeed et al.

Saeed et al. BMC Complementary and Alternative Medicine 2012 12:221, doi: 10.1186/1472-6882-12-221

Superoxide Radical Scavenging Activity

Superoxide radical is considered a major biological source of reactive oxygen species [34]. Although superoxide anion is a weak oxidant, it gives rise to generation of powerful and dangerous hydroxyl radicals as well as singlet oxygen, both of which contribute to oxidative stress [35]. The superoxide radical scavenging effect of different fractions was compared with the same doses of ascorbic acid ranging from 25 - 500 μg/ml. The EC_{50} values in superoxide scavenging activities were in the order of TLM > TLC > TLB > TLE > TLA > TLH (Table 1). When compared to ascorbic acid; the superoxide scavenging activity of the extract was found to be low ($P < 0.05$). In spite of this TLM and TLC (EC_{50} 55.0±0.3 and 68.0±2 μg/ml respectively) behave as powerful superoxide anion scavengers that may include therapeutic use against oxidative stress.

Phosphomolybdate Assay

The phosphomolybdate method is quantitative, since the total antioxidant capacity (TAC) is expressed as ascorbic acid equivalents. The antioxidant capacity of various solvent fractions of *T.leptophylla* was found to decrease in this order: TLB > TLC > TLE > TLM > TLA > TLH fraction (Table 1; Figure 1C). All results showed antioxidant activity in dose dependent manner at concentration 25 to 250 μg/ml. The EC_{50} value of antioxidant capacity for the TLB (10.7±2 μg/ml) was most pronounced ($P < 0.05$) than TLC (135±4 μg/ml) and TLE (196±4 μg/ml) (Table 1). Strong antioxidant activity of TLB statistically similar to ascorbic acid indicates strong antioxidants in this fraction and these could be attributable to the presence of phenolic compounds.

Hydroxyl Radical Scavenging Activity

Hydroxyl radical scavenging activity was quantified by measuring the inhibition of the degradation of 2-deoxyribose by the free radicals generated by the Fenton reaction. The hydroxyl radical scavenging activity of *T. leptophylla* extract and its derived fractions can be ranked as TLC > TLE > TLB > TLM > TLH and TLA (Table 1). All fractions showed antioxidant activity in dose dependent manner at concentration 25 - 500 μg/ml (Figure 1D). In the present investigation, the EC_{50} value of hydroxyl radical scavenging activity for the TLC and TLE was 8.0±1 and

27.0±1 μg/ml while for TLB was 62.0±2 μg/ml (Table 1). The markedly strong ($P < 0.05$) antioxidant response of TLC and TLE in comparison with ascorbic acid might be helpful in characterizing the significant sources of natural antioxidant reaction.

Hydrogen Peroxide Radical Scavenging Activity

The scavenging effect of different fractions of *T. leptophylla* on hydrogen peroxide was concentration-dependent (25 - 500 μg/ml) as shown in Figure 1E. TLE displayed strong H_2O_2scavenging activity (EC_{50} 68.0±2 μg/ml) whereas that of the standard, ascorbic acid exhibited 9.0±0.3 μg /ml. The scavenging activities of TLC and TLM were (EC_{50} 122±5 μg/ml and 130±5 μg/ml, respectively) (Table 1). EC_{50} values of the fractions in scavenging hydrogen peroxide were significantly different ($P < 0.05$) from the EC_{50} values obtained for ascorbic acid. The scavenging activity for hydrogen peroxide of various solvent extracts from *T. leptophylla* was in the order of TLE > TLC > TLM > TLH > TLB > TLA respectively.

ABTS Radical Scavenging Activity

All the fractions of *T. leptophylla* scavenged ABTS radical in a concentration-dependent way (25 - 250 μg/ml) (Figure 1F). Present results showed that the ABTS radical scavenging ability of samples can be ranked as TLB > TLA > TLE > TLH > TLC > TLM. TLB and TLA exhibited prominent ABTS radical scavenging activities.

Reducing Power Activity

Figure 2 shows the dose response curves for the reducing powers of all extracts (25 - 250 μg/ml) from *T. leptophylla*. It was found that the reducing power increased with concentration of each sample. The ranking order for reducing power was TLM > TLB > TLC > TLA > TLE > TLH. Significantly higher reducing power (1.47±0.14 at 250 μg/ml) was evident in TLM fraction.

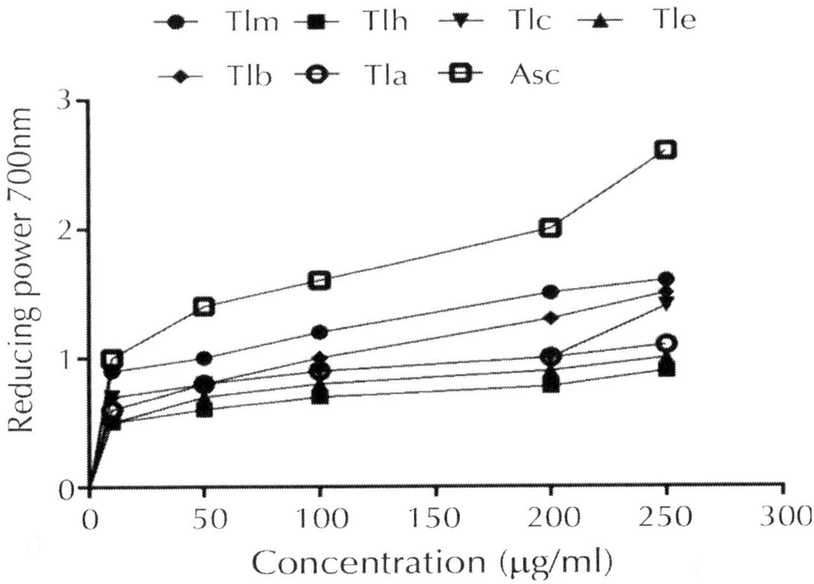

Figure 2: Reducing power of different extracts from the methanol extract of *T. leptophylla* by different solvents at different concentrations. Each value represents a mean ± SD (n = 3). Tlm., methanolic fraction, Tlh., *n*-hexane fraction, Tlc., chloroform fraction, Tle., ethyl acetate fraction, Tlb., *n*-butanol fraction, Tla., residual aqueous fraction, Asc., ascorbic acid.

Extraction Yield, Total Phenolic and Flavonoid Contents

The extraction yield of different fractions of *T. leptophylla* varied from 4.3±0.4 to 11.5±1.6% with a descending order of TLM > TLA > TLE > TLH > TLB > TLC (Table 2). So the extraction with methanol resulted in the highest amount of total extractable compounds whereas the extraction yield with chloroform was only small in comparison with that of the other solvents.

Table 2: Total phenolics, flavonoid and extraction yield of methanol extract and soluble fractions of *T. leptophylla*

Plant extracts	Total phenolics (mg gallic acid equivalent/g)	Total flavonoid (mg rutin equivalent/g)	Extraction yield (%)
Methanol extract	121.9±3.1c	59.6±1.5c	11.5±1.6d
n-Hexane fraction	54.9±2.8a	15.8±0.9a	5.4±1.3b
Chloroform fraction	78.0±1.1b	26.0±1.9b	4.3±0.4a
Ethyl acetate fraction	80.0±5.2b	60.9±2.2c	6.1±0.8b
n-Butanol fraction	80.9±2.9b	55.0±2.5c	4.8±1.1a
Residual aqueous fraction	49.9±4.1a	14.9±0.4a	8.2±0.9c

Each value in the table is represented as mean ± SD (n = 3). Values in the same column followed by a different letter (a-d) are significantly different ($P < 0.05$).

Saeed et al.

Saeed et al. BMC Complementary and Alternative Medicine 2012 12:221, doi:10.1186/1472-6882-12-221

Total phenolic content was estimated by using Folin-Ciocalteu reagent. Total phenolic content of the different fractions of *T. leptophylla* was solvent dependent and expressed as milligrams of gallic acid equivalents (GAE) equivalent. Table 2 summarizes that total phenolic compounds in fractions varied widely, ranging from 49.9±4.1 and 121.9±3.1 mg/g expressed as gallic acid equivalents (GAE). TLM exhibited the highest total phenolic content. The content of flavonoid expressed as rutin equivalents, varied from 14.9±0.4 to 60.9±2.2 mg rutin equivalent/g extract (Table 2). The TLE showed the highest amount of flavonoid contents followed by TLM and TLB.

Correlation of EC_{50} Values of Antioxidant Activities with TPC and TFC

For total phenolic content a significant but marginal positive correlation ($R^2 > 0.7020$, $R^2 > 0.7458$, $R^2 > 0.6663$ and $R^2 > 0.6777$ respectively) was found between TPC and EC_{50} values for DPPH, hydroxyl, phosphomolybdate and ABTS respectively, while also another weak positive correlation was found between TFC and EC_{50} values for superoxide anion and hydroxyl radicals. However, a non significant correlation was found in case of hydrogen peroxide radical scavenging with both TPC and TFC (Table 3).

Table 3: Correlations between the EC50 values of antioxidant activities and phenolic and flavonoids content of *T. leptophylla*

Assays	Correlation R2	
	Phenolics	Flavonoid
EC50 of DPPH radical scavenging ability	0.7020*	0.192
EC50 of superoxide radical scavenging ability	0.1905	0.6613*
EC50 of phosphomolybdate antioxidant capacity	0.6663*	0.4306
EC50 of hydroxyl radical scavenging ability	0.7458*	0.7215*
EC50 of hydrogen peroxide radical scavenging ability	0.0194	0.0126
EC50 of ABTS radical scavenging ability	0.6777*	0.1285

Each value in the table is represented as mean ± SD (n = 3). * indicates significance at $P < 0.05$.

Saeed et al.

Saeed et al. BMC Complementary and Alternative Medicine 2012 12:221, doi:10.1186/1472-6882-12-221

In Vivo Antioxidant Activity

In vivo lipid peroxidation study reveals that rats of CCl_4 treated group showed significant increase ($P < 0.05$) in thiobarbituric acid reactive substances (TBARS) when compared with rats of normal control group. TLM and silymarin were able to significantly blunt ($P < 0.05$) this rise in TBARS level (Table 4). There was a marked decrease in the level of GSH in CCl_4 treated group when compared with normal control group. The GSH level was significantly increased ($P < 0.05$) in TLM and silymarin treated groups (Table 4).

Table 4: Effects of TLM on hepatic protein, lipid peroxidation (TBARS) and glutathione (GSH)

Group	Protein (μg/mg tissue)	TBARS (nM/mg protein)	GSH (μg/mg protein)
1	1.697±0.024+	3.43 ± 0.53+	31.34 ± 1.10+
II	0.657±0.021*	6.39 ± 0.98*	15.99 ± 1.67*
III	1.563±0.092*+	4.30 ± 0.91*+	26.68 ± 1.16*+
IV	1.509±0.003+	4.69 ± 0.74*+	27.42 ± 1.21*+
V	1.702±0.012+	3.67 ± 0.42+	29.97 ± 1.12+

Group I: Controls (Olive oil+ DMSO); II: CCl_4 (3 ml/kg bw); III: TLM (200 mg/kg bw) +CCl_4 (3 ml/kg bw); IV: Silymarin (50 mg/kg bw) +CCl_4 (3 ml/kg bw); V: TLM (200 mg/kg bw).

TLM; methanol extract of *T. leptophylla*.

Values are Mean ± SD (06 numbers).

*, Indicate significance from control group ($P < 0.05$).

+, Indicate significance from CCl_4 group ($P < 0.05$).

Saeed *et al.*

Saeed *et al. BMC Complementary and Alternative Medicine* 2012 **12**:221 doi: 10.1186/1472-6882-12-221

DISCUSSION

Several techniques have been used to determine the antioxidant activity *in vitro* in order to allow rapid screening of substances since substances that have low antioxidant activity *in vitro*, will probably show little activity *in vivo*[8]. Free radicals are known to play a definite role in a wide variety of pathological manifestations. Antioxidants fight against free radicals and protect us from various diseases. They exert their action either by scavenging the reactive oxygen species or protecting the antioxidant defense mechanisms [21].

The electron donation ability of natural products can be measured by 2,2'-diphenyl-1- picrylhydrazyl radical (DPPH) purple-coloured solution bleaching [8]. The method is based on scavenging of DPPH through the addition of a radical species or antioxidant that decolourizes the DPPH solution. The degree of colour change is proportional to the concentration and potency of the antioxidants. A large decrease in the absorbance of the reaction mixture indicates significant free radical scavenging activity of the compound under test [36]. In the present study among all the fractions tested, *n*-butanol, chloroform and ethyl acetate showed significantly higher inhibition percentage and positively correlated with total phenolic content. Results of this study suggest that the plant extract contain phytochemical constituents that are capable of donating hydrogen to a free radical to scavenge the potential damage.

Superoxide radical is considered a major biological source of reactive oxygen species [34]. Although superoxide anion is a weak oxidant, it gives rise to generation of powerful and dangerous hydroxyl radicals as well as singlet oxygen, both of which contribute to oxidative stress [35]. The results of our study revealed that TLM, TLC and TLB had effective capacity of scavenging for superoxide radical and correlated with total flavonoid content thus suggesting its antioxidant potential.

The antioxidant capacity of the fractions was measured spectrophotometrically through phosphomolybdenum method, based on the reduction of Mo (VI) to Mo (V) by the test sample and the subsequent formation of green phosphate/Mo (V) compounds with a maximum absorption at 765 nm. The present study demonstrated that TLB exhibited the highest antioxidant capacity for phosphomolybdate reduction. Recent studies have shown that many flavonoid and related polyphenols contribute significantly to the phosphomolybdate scavenging activity of medicinal plants [37,38].

Hydroxyl radical is one of the potent reactive oxygen species in the biological system. It reacts with polyunsaturated fatty acid moieties of cell membrane phospholipids and causes damage to cell [5,39]. The hydroxyl radical is regarded as a detrimental species in pathophysiological processes and capable of damaging almost every molecule of biological system and contributes to carcinogenesis, mutagenesis and cytotoxicity [40]. Hydroxyl radicals were produced by the reaction of H_2O_2 and the ferrous that would react with 2-deoxyribose. The reaction was stopped by adding TBA reagent that would give a red colour if the malonaldehyde was formed as the result of the reaction between the radical and 2-deoxyribose. Hydroxyl radical scavenging capacity of an extract is directly proportional to its antioxidant activity which is depicted by the low intensity of red colour [41]. All fractions of *T. leptophylla* when added to the reaction mixture actively scavenged the hydroxyl radicals and prevented the degradation of 2-deoxyribose.

Hydrogen peroxide occurs naturally at low concentration levels in the air, water, human body, plants, microorganisms and food [41]. H_2O_2 is rapidly decomposed into oxygen and water and this may produce hydroxyl radicals (·OH) that can initiate lipid peroxidation and cause DNA damage[42]. Ethyl acetate fraction of *T. leptophylla* efficiently scavenged hydrogen peroxide which may be attributed to the presence of phenolic groups that could donate electrons to hydrogen peroxide, thereby neutralizing it into water.

ABTS radical scavenging assay involves a method that generates a blue/green $ABTS^{+}$chromophore via the reaction of ABTS and potassium persulfate. The ABTS radical cation is generated by the oxidation of ABTS with potassium persulfate, its reduction in the presence of hydrogen-donating antioxidants is measured spectrophotometrically at 745 nm. All the fractions possessed strong ABTS scavenging activity an observation that is supported by other researchers[43].

In reducing power assay, the yellow colour of the test solution changes to green depending on the reducing power of the test specimen. The presence of the reductants in the solution causes the reduction of the Fe^{3+}/ferricyanide complex to the ferrous form. Therefore, Fe^{2+} can be monitored by absorbance measurement at 700 nm. Previous reports suggested that the reducing properties have been shown to exert antioxidant action by donating of a hydrogen atom to break the free radical chain [44]. Increasing absorbance at 700 nm indicates an increase in reducing ability. The antioxidants present in the fractions of *T. leptophylla* caused their reduction of Fe^{3}+/ ferricyanide complex to the ferrous form, and thus proved the reducing power.

Plant materials rich in phenolics are increasingly being used in the food industry because they retard oxidative degradation of lipids and improve the quality and nutritional value of food [45].

Phenolic compounds are considered secondary metabolites and these phytochemical compounds derived from phenylalanine and tyrosine occur ubiquitously in plants and are diversified [46]. The methanol extract exhibited the highest total phenolics content, whereas the contents obtained with residual aqueous fraction were much smaller that is in agreement with other reports [47]. Phenolic compounds of plants are also very important because their hydroxyl groups confer scavenging ability.

Phenolic compounds of plants fall into several categories; chief among these are the flavonoids which have potent antioxidant activities [8]. Flavonoids are naturally occurring in plants and are thought to have positive effects on human health. Studies on flavonoidic derivatives have shown a wide range of antibacterial, antiviral, anti inflammatory, anticancer, and anti-allergic activities[48,49]. Flavonoids have been shown to be highly effective scavengers of most oxidizing molecules, including singlet oxygen, and various free radicals [50] implicated in several diseases. So comparable with the findings in the literature for other extracts of plant products [42] our results suggested that phenolic acids and flavonoids may be the major contributors for the antioxidant activity as the EC_{50} values of radical scavenging activity of various soluble fractions of*T. leptophylla* and the contents of phenolics or flavonoids exhibited significant correlation. However, non significant correlation was found in case of hydrogen peroxide radical scavenging. It is known that different phenolic compounds have different responses in the Folin-Ciocalteu method. Similarly the molecular antioxidant response of phenolic compounds varies remarkably, depending on their chemical structure [51]. In addition, there may be some interference rising from other chemical components present in the extract, such as sugars or ascorbic acid [52].

Liver damage is very common since liver has to detoxicate a lot many toxic substances. There are several chemicals that have been known to induce hepatotoxicity by producing the reactive species which form covalent bonds with the lipids of the tissue [53,54]. Liver injury due to CCl_4 in rats was first reported in 1936 [55] and has been widely and successfully used by many investigators [56,57]. Carbon tetrachloride is metabolized by cytochrome P-450 in endoplasmic reticulum and mitochondria with the formation of CCl_3O^-, a reactive oxidative free radical, which initiates lipid peroxidation [58,59].

A large reserve of reduced glutathione is present in hepatocytes for detoxification of free radicals. However, oxidative stress results in toxicity when the rate at which the ROS are generated exceeds the cell capacity for their removal [60,61]. Most hepatotoxic chemicals damage liver by inducing, directly or indirectly, lipid peroxidation. MDA is one of the end products in the lipid peroxidation process [62]. In our *in vivo* study elevation in levels of end products of lipid peroxidation in liver of rats treated with CCl_4 were observed. The increase in MDA levels in liver suggests enhanced lipid peroxidation leading to tissue damage [63]. Treatment with TLM significantly reversed these changes. Hence it may be possible that the mechanism of hepatoprotection of extract is due to its antioxidant effect [63].

GSH is an intracellular reductant and protects cells against free radicals, peroxides and other toxic compounds. GSH is a naturally occurring substance that is abundant in many living creatures; GSH depletion increases the sensitivity of cells to various aggressions leading to tissue disorder and injury [64]. In the present study we demonstrated the effectiveness of the extract by using CCl_4induced rats and found that exogenous TLM supplementation elevated GSH levels in rats with CCl_4treatment and thus might provide a mean of recovering reduced GSH levels and to prevent tissue disorders and injuries. Therefore, it is valid to consider that TLM, because of its antioxidant property as the extract has confirmed the presence of saponins, might be capable of protecting the hepatic tissue from CCl_4-induced injury and inflammatory changes. Saponins are natural products, which have been shown to possess antioxidant properties [65,66].

CONCLUSIONS

The replacement of synthetic with natural antioxidants (because of implications for human health) may be advantageous. In the present study analysis of free radical scavenging activity and total phenolic and flavonoid content showed that mainly the TLB, TLC, TLE and TLM from the whole plant of *T. leptophylla* can be the potent source of natural antioxidants. The results of *in vivo* studies suggest that methanolic extract of *T. leptophylla* may be useful in defense against CCl_4-induced liver damage possibly be due to its antioxidant properties.

AUTHORS' CONTRIBUTIONS

NS made significant contribution to acquisition of data, analysis, drafting of the manuscript. MRK has made substantial contribution to conception and design, interpretation of data, drafting and revising the manuscript for intellectual content. MS participated in the design and collection of data and analysis. All authors read and approved the final manuscript.

REFERENCES

1. Maqsood S, Singh P, Samoon MH, Balange AK: Effect of dietary chitosan on non-specific immune response and growth of Cyprinus carpio challenged with Aeromonas hydrophila. *Inter Aqua Res* 2010, 2:77-85.
2. Zengin G, Cakmak YS, Guler GO, Aktumsek A: Antioxidant properties of methanolic extract and fatty acid composition of Centaurea urvillei DC. subsp. hayekiana Wagenitz. *Rec Nat Prod* 2011, 5:123-132.
3. Vongtau HO, Abbah J, Chindo BA, Mosugu O, Salawu AO, Kwanashie HO, Gamaniel KS:Central inhibitory effects of the methanol extract of Neorautanenia mitis root in rats and mice. *J Pharm Biol* 2005, 43:113-120.
4. Oluyemi KA, Okwuonu UC, Baxter DG, Oyesola TO: Toxic effects of methanolic exract of Aspilia africana leaf on the estrous cycle and uterine tissues of Wistar rats. *Int J Morphol* 2007, 25:609-614.
5. Halliwell B, Gutteridge JMC: Formation of thiobarbituric acid reactive substances from deoxyribose in the presence of iron salts: the role of superoxide and hydroxyl radicals. *FEBS Lett* 1981, 128:347-352.
6. Gulcin I: Antioxidant activity of food constituents: an overview. *Arch Toxicol* 2012, 86:345-391.
7. Gocer H, Gulcin I: Caffeic acid phenethyl ester (CAPE): correlation of structure and antioxidant properties. *Int J Food Sci Nutr* 2011, 62:821-825.

8. Nunes PX, Silva SF, Guedes RJ, Almeida S: *Biological oxidations and antioxidant activity of natural products, Phytochemicals as nutraceuticals - Global Approaches to Their Role in Nutrition and Health*. 2012.

9. Djeridane A, Yousfi M, Nadjemi B, Boutassouna D, Stocker P, Vidal N: Antioxidant activity of some Algerian medicinal plants extracts containing phenolic compounds. *Food Chem* 2006, 97:654-660.

10. Wannes WA, Mhamdi B, Sriti J, Jemia MB, Ouchikh O, Hamdaoui G, Kchouk ME, Marzouk B:Antioxidant activities of the essential oil and methanol extracts from myrtle (Myrtus communis var. italica L.) leaf, stem and flower. *Food Chem Toxicol* 2010, 48:1362-1370.

11. Harman D: *Free radical theory of aging. Current status*. Amsterdam: Elsevier; 1998:p. 3-p. 7.

12. Osawa T, Kavakishi S, Namiki M, Kuroda Y, Shankal DM, Waters MD: *Antimutagenesis and anticarcinogenesis mechanisms II*. New York: Plenum; 1990:139-153.

13. Govind P: Medicinal plants against liver diseases. *IJPR* 2011, 2:115-121.

14. Baranski R, Baranska M, Schulz H, Simon PW, Nothnaged T: Single seed roman measurements allow taxonomical discrimination of Apiaceae accessions collected in gene banks. *Biopolymers* 2006, 81:497-505.

15. Bigdeli M, Rustaiyan A, Masoudi S: Composition of the essential oil of Torilis arvensis (Huds). Link from Iran. *J Essential Oil Res* 2004, 16:526-527.

16. Maleki S, Seyyednejad SM, Damabi NM, Motamedi H: Antibacterial activity of the fruits of Iranian Torilis leptophylla against some clinical pathogens. *Pak J Biol Sci* 2008, 11:1286-1289.

17. Kil HY, Seong ES, Ghimire BK, Chung IM, Kwon SS, Goh EJ, Hoe K, Kim MJ, Lim JD, Lee D, Yu CY: Antioxidant and antimicrobial activities of crude sorghum extract. *Food Chem* 2009, 115:1234-1239.

18. Brand-Williams W, Cuvelier ME, Berset C: Use of free radical method to evaluate antioxidant activity. *Lebensmittel-Wissenschaft und-Technol* 1995, 28:25-30.

19. Bursal E, Gulcin I: Polyphenol contents and in vitro antioxidant activities of lyophilized aqueous extract of kiwifruit (Actinidia deliciosa). *Food Res Int* 2011, 44:1482-1489.

20. Beauchamp C, Fridovich I: Superoxide dismutase: improved assays and an assay applicable to acrylamide gels. *Anal Biochem* 1971, 44:276-277.

21. Umamaheswari M, Chatterjee TK: In vitro antioxidant activities of the fractions of Coccinnia grandis L. leaf extract. *Afr J Trad Compl Altern Med* 2008, 5:61-73.

22. Ilavarasan R, Mallika M, Venkataraman S: Anti-inflammation and antioxidant activities of Cassia fistula Linn. bark extracts. *Afr J Trad Compl Altern Med* 2005, 2:70-85.

23. Ruch RJ, Cheng S-J, Klaunig JE: Prevention of cytotoxicity and inhibition of intercellular communication by antioxidant catechins isolated from Chinese green tea. *Carcinogenesis* 1989, 10:1003-1008.

24. Re R, Pellegrini N, Proteggente A, Pannala A, Yang M, Rice-Evans C: Antioxidant activity applying an improved ABTS radical cation decolourisation assay. *Free Rad Biol Med* 1999, 26:1231-1237.

25. Gulcin I, Topal F, Cakmakc R, Bilsel M, Goren AC, Erdogan U: Pomological features, nutritional quality, polyphenol content analysis and antioxidant properties of domesticated and three wild ecotype forms of raspberries (Rubus idaeus L.). *J Food Sci* 2011, 76:585-593.

26. Fejes S, Blazovics A, Lugasi A, Lemberkovics E, Petri G, Kery A: In vitro antioxidant activity of Anthriscus cerefolium L. (Hoffm.) extracts. *J Ethnopharmacol* 2000, 69:259-265.

27. Kim DO, Jeong SW, Lee CY: Antioxidant capacity of phenolic phytochemicals from various cultivars of plums. *Food Chem* 2003, 81:321-326.

28. Park Y-S, Jung S-T, Kang S-G, Heo BK, Arancibia-Avila P, Toledo F, Drzewiecki J, Namiesnik J, Gorinstein S: Antioxidants and proteins in ethylene-treated kiwifruits. *Food Chem* 2008, 107:640-648.

29. Harborne JB: *Phytochemical methods*. London: Chapman and Hall, Ltd; 1973:pp. 49-pp. 188.

30. Trease GE, Evans WC: *Pharmacognsy*. 11th edition. London: Brailliar Tiridel Can Macmillian Publishers; 1989:60-75.

31. Sofowara AE: *Medicinal plants and traditional medicine in Africa*. 2nd edition. Ibadan, Nigeria: Spectrum books Ltd.; 1993:289.
32. Iqbal M, Sharma SD, Zadeh HR, Hasan N, Abdulla M, Athar M: Glutathione metabolizing enzymes and oxidative stress in ferric nitrilotriacetate (Fe- NTA) mediated hepatic injury. *Redox Rep* 1996, 2:385-391.
33. Jollow DJ, Mitchell JR, Zampaglione N, Gillete JR: Bromobenzene induced liver necrosis. Protective role of glutathione and evidence for 3, 4- bromobenzene oxide as a hepatotoxic metabolite. *Pharmacology* 1974, 1:151-169.
34. Alves CQ, David JM, David JP, Bahia MV, Aguiar RM: Methods for determination of in vitro antioxidant activity for extracts and organic compounds. *Química Nova* 2010, 33:2202-2210.
35. Meyer AS, Isaksen A: Application of enzymes as food antioxidants. *Trends Food Sci Tech* 1995, 1995(6):300-304.
36. Krishnaiah D, Sarbatly R, Nithyanandam RR: A review of the antioxidant potential of medicinal plant species. *Food Bioprod Process* 2011, 89:217-233.
37. Sharififar F, Dehghn-Nudeh G, Mirtajaldini M: Major flavonoids with antioxidant activity from Teucrium polium L. *Food Chem* 2009, 112:885-888.
38. Khan RA, Khan MR, Sahreen S: Assessment of flavonoids contents and in vitro antioxidant activity of Launaea procumbens. *Chem Central J* 2012, 6:43.
39. Khan RA, Khan MR, Sahreen S, Ahmed M: Evaluation of phenolic contents and antioxidant activity of various solvent extracts of Sonchus asper (L.) Hill. *Chem Central J* 2012, 6:12.
40. Babu BH, Shylesh BS, Padikkala J: Antioxidant and hepatoprotective effect of Alanthus icicifocus. *Fitoterapia* 2001, 72:272-277
41. Gulcin I, Berashvili D, Gepdiremen A: Antiradical and antioxidant activity of total anthocyanins from Perilla pankinensis decne. *J Ethnopharmacol* 2005, 101:287-293.
42. Sahreen S, Khan MR, Khan RA: Phenolic compounds and antioxidant activities of Rumex hastatus D. Don. Leaves. *J Med Plants Res* 2011, 5:2755-2765.

43. Sahreen S, Khan MR, Khan RA: Evaluation of antioxidant activities of various solvent extracts of Carissa opaca fruits. *Food Chem* 2010, 122:1205-1211.

44. Gordon MH: The mechanism of antioxidant action in vitro. In *Food antioxidants*. Edited by Hudson BJ. London: Elsevier Applied Science; 1990:pp. 1-pp. 18.

45. Kähkönen MP, Hopia AI, Vuorela HJ, Rauha JP, Pihlaja K, Kujala TS, Heinonen M:Antioxidant activity of plant extracts containing phenolic compounds. *J Agri Food Chem* 1999, 1999(47):3954-3962.

46. Naczk M, Shahidi F: Extraction and analysis of phenolics in food. *J Chromatogr A* 2004, 1054:95-111.

47. Ao C, Li A, Elzaawely AA, Xuan DT, Twata S: Evaluation of antioxidant and antibacterial activities of Ficus microcarpa L. fill extract. *Food Contrl* 2008, 19:940-948.

48. Di Carlo G, Mascolo N, Izzo AA, Capasso F: Flavonoids: old and new aspects of a class of natural therapeutic drugs. *Life Sci* 1999, 65:337-353.

49. Montoro P, Braca A, Pizza C, De Tommasi N: Structure-antioxidant activity relationships of flavonoids isolated from different plant species. *Food Chem* 2005, 92:349-355.

50. Bravo L: Polyphenols: chemistry, dietary sources, metabolism and nutritional significance. *Nutr Reviews* 1998, 56:317-333.

51. Satue-Gracia MT, Heinonen M, Frankel EN: Antioxidant activity of anthocyanin in LDL and lecithin liposome systems. *J Agric Food Chem* 1997, 1997(45):3362-3367.

52. Singleton VL, Rossi JA: Colourimetry of total phenolics with phosphomolybdic-phosphotungstic acid reagents. *Am J Enol Vitic* 1965, 16:144-158.

53. Deshwal N, Sharma1 AK, Sharma P: Review on hepatoprotective plants. *IJPSR* 2011, 7:15-26.

54. Amartya GK, Partha G, Upal MK, Shibnath G: Hepatoprotective & antioxidant effect & stereoidal saponins of solanum of Solanum xanthocarpum and Solanum nigrum in paracetomol induce hepatotoxicity in rats. *Pharmacologyonline* 2009, 1:757-768.

55. Cameron GR, Thomas JC, Karunarathe WAE: The pathogenesis of liver injury in carbon tetrachloride and thioacetamide poisioning. *J Path Bact* 1936, 41:297.

56. Handa SS, Sharma A: Hepatoprotective activity of andrographolide from Andrographis paniculata against carbon tetrachloride. *Ind J Med Res* 1990, 92:276-292.

57. Shirwaiker A, Sreenivasan KK, Krishnanand BR, Kumar AV: Chemical investigation and anti hepatotoxic activity of the root bark of Caparis spinos. *Fitoterapia* 1996, 67:200-204.

58. Zimmerman MD, Hayman J: *Function and integrity of the liver*. 17th edition. 1976, 217-250. [*In: Clinical diagnosis and management by laboratory methods In: Clinical diagnosis and management by laboratory methods*]

59. Agarwal AK, Mehendale JK: Potentiation of carbon tetrachloride hepatotoxicity and lethality by chlordecone in female rats. *Toxicology* 1983, 26:231-242.

60. Ghosh T, Maity TK, Das M, Bose A, Dash D: In vitro antioxidant and hepatoprotective activity of ethanolic extract of Bacopa monnieri L. aerial parts. *IJPT* 2007, 6:77-85.

61. Kurata M, Suzuki M, Agar NS: Antioxidant systems and erythrocyte life span in mammals. *Biochem Physiol* 1993, 106:477-487.

62. Asha VV: Preliminary studies on hepatoprotective activities of Momordica sabangulata and Naragama alat. *Indian J Pharmacol* 2001, 33:276-279.

63. Khan MR, Marium A, Shabbir M, Saeed N, Bokhari J: Antioxidant and hepatoprotective effects of Oxalis corniculata against carbon tetrachloride (CCl4) induced injuries in rat. *Afr J Pharm Pharmacol* 2012, 6:2255-2267.

64. Jollow DJ: Glutathione thresholds in reactive metabolite toxicity. *Arch Toxicol Suppl* 1980, 3:95-110.

65. Yoshiki Y, Okubo K: Active oxygen scavenging activity of DDMP (2, 3-dihydro-2, 5-dihydroxy-6-methyl-4H-pyran-4-one) saponin in soybean seed. *Biosci Biotech Biochem* 1995, 59:56-57.

66. Hu J, Lee SO, Hendrich S, Murphy PA: Quantification of the group B soyasaponins by high-performance liquid chromatography. *J Agri Food Chem* 2002, 50:87-94.

Chapter 6

The Potential of Rice to Offer Solutions for Malnutrition and Chronic Diseases

Sharifa Sultana Dipti[1,12], Christine Bergman[12,2], Siti Dewi Indrasari[12,3], Theja Herath[12,4], Robert Hall[12,5,6], Hueihong Lee[12,7], Fatemeh Habibi[12,8], Priscila Zaczuk Bassinello[12,9], Eduardo Graterol[10,12], Julie P Ferraz[11], and Melissa Fitzgerald[1,12,13,14]

[1]Grain Quality and Nutrition Centre, International Rice Research Institute (IRRI), DAPO 7777, Metro Manila, Philippines

[2]Department of Food and Beverage, University of Nevada-Las Vegas, Las Vegas, NV, 89154, USA

[3]Indonesian Center for Rice Research (ICRR) BB Padi, Jl. Raya 9, Sukamandi, Subang, 41256, Jawa Barat, Indonesia

[4]Industrial Technology Institute, Colombo 7, Bauddhaloka Mawatha, 363, Sri Lanka

[5]Plant Research International, PO Box 98, 6700AB, Wageningen, The Netherlands

[6]Centre for BioSystems Genomics, P.O. Box 98, 6700AB, Wageningen, The Netherlands

[7]Faculty of Agriculture and Food Sciences, Universiti Putra Malaysia, Nyabau Road, 97000, Bintulu Sarawak, Malaysia

[8]Rice Research Institute of Iran (RRII), Km5 Tehran Rd, 41996-13475, Rasht,I.R, Iran

[9]Embrapa Arroz e Feijão, Rodovia GO-462, Km 12, Zona Rural C.P. 179, Santo Antônio de Goiás, GO, 75375-000, Brazil

[10]Fundación para la Investigación Agrícola DANAC, Apartado Postal 182, San Felipe, Estado Yaracuy, Venezuela

[11]Institute of Science, Diabetes Foundation Marikina, Philippines, Healthserve Hospital, and Calamba Doctors Hospital, Laguna, Philippines

[12]International Network for Quality Rice, Metro Manila, Philippines

[13]Grain Quality and Nutrition Centre, International Rice Research Institute (IRRI), DAPO 7777, Metro Manila, Philippines

[14]Current address: School of Agriculture and Food Science, University of Queensland, St Lucia, 4072, Australia

ABSTRACT

It is internationally accepted that malnutrition and chronic diseases in developing countries are key limitations to achieving the Millennium Development Goals. In many developing countries, rice is the primary source of nutrition. In those countries, the major forms of malnutrition are Fe-induced anaemia, Zn deficiency and Vitamin A deficiency, whereas the major chronic disease challenges are Type II diabetes, cardiovascular disease and some cancers. There is a growing corpus of evidence regarding both limitations and opportunities as to how rice could be an effective vehicle by which to tackle key nutrition and health related problems in countries with limited resources. Rice breeding programs are able to focus on developing new varieties carrying enhanced amounts of either Fe, Zn or beta-carotene because of large public investment, and the intuitive link between providing a mineral/vitamin to cure a deficiency in that mineral/vitamin. By contrast, there has been little investment in progressing the development of particular varieties for potential impact on chronic diseases. In this review article we focus on the broad battery of evidence linking rice-related nutritional limitations to their impact on a variety of human health issues. We discuss how rice might offer sometimes even simple solutions to rectifying key problems through targeted biofortification strategies and finally, we draw attention to how recent technological (−omics) developments may facilitate untold new opportunities for more rapidly generating improved rice varieties specifically designed to meet the current and future nutritional needs of a rapidly expanding global population.

REVIEW

Background

As we enter the second decade of the 21st century, experts agree that the world faces three major global health challenges. The first is completing the work to meet the millennium development targets of decreased malnutrition and infectious disease. The second is the alarming increase in the incidence of chronic diseases like heart disease, Type II diabetes, obesity, and cancers in developing nations. The third is a consequence of globalisation whereby traditional diets are being replaced or supplemented with nutritionally compromised fast-foods. Nutrition is a feature of each of these challenges, and in a sad twist of irony, developing countries, where rice is the staple, are the hardest hit by all three of these global challenges.

Over the past decade, rice improvement programs have included key nutritional targets in their breeding programs, attempting to meet specific targets for Fe, Zn and pro-vitamin A content. During this time, significant progress has been made in understanding both ways to increase the micronutrient content of the polished and unpolished rice, and the limitations to achieving those targets in conventional breeding programs. However, the potential of rice to contribute to the prevention or management of chronic diseases is not so widely recognised, and research aimed to quantify that potential receives a fraction of the public funding of rice and malnutrition, despite the massive and growing problem of chronic disease that prevails in all rice-consuming countries ([Nugent 2008]). Perhaps this is because the association between compounds in rice grains and chronic diseases is more complex and less intuitive than, for example, increasing Zn content of grains to address Zn deficiency. A second reason could be limitations in the detection and identification of relevant grain constituents, meaning that phenotyping tools are not available for breeding objectives. The previous decade has borne witness to breath-taking technological advances, which should enable research to progress more rapidly both in identifying grain constituents, and determining the impact of these in the prevention or management of chronic diseases.

This review will focus on the nutritional potential of grains of both polished and unpolished rice, for the most pressing issues of malnutrition and chronic diseases in rice-consuming countries. We discuss opportunities and obstacles, and identify roles that rice might play in health and nutritional impact, and patterns of rice consumption that could contribute to solutions for the grand challenges to global health.

MALNUTRITION

Malnutrition in Rice-Consuming Populations

Iron deficiency anaemia is a worldwide public health problem, with global prevalence estimated at 24.8% (95% CI: 22.9–26.7) ([Shaw and Friedman 2011]). It occurs when the concentration of haemoglobin (Hb) falls below 11 g/dl in pregnant women, 12 g/dl in non-pregnant women aged 15–49, and 11 g/dl in children under five. Anaemia can cause maternal mortality associated with childbirth. In adults it lowers work performance, and it has been linked with reduced immune competence ([Shaw and Friedman 2011]). The majority of the disease burden is shouldered by developing countries with high levels of rice consumption. The highest prevalence is found in Africa, the Middle East, Central, South and South-East Asia, and areas of Latin America, where two thirds of children under five, and almost 50% of women are anaemic ([WHO 2008]). One reason that Fe-deficient anaemia is widespread amongst rice-consuming countries is because of the low concentration of Fe in polished rice, like other starchy staples, combined with the inability of poor people to supplement the staple with other foods rich in micronutrients. A survey of 56 varieties showed that the average Fe content of the polished grains was 4.3

ppm ([Bounphanousay 2007]); however, the biological availability of Fe from polished rice is low ([Shaw and Friedman 2011]).

The first cases of zinc deficiency were described in the Middle East the 1960s, and attributed to the consumption of diets high in anti-nutritional factors ([Prasad et al. 1963]). Zn deficiency is now recognised as one of the five major factors contributing to disease burden in developing countries ([WHO 2002]). Zn deficiency leads to decreased neuropsychological function, it contributes to childhood mortality by increasing the incidence and severity of acute and chronic diarrhoea, and in pregnant women it leads to difficulties in childbirth, retarded foetal growth, and foetal abnormalities ([Prasad 2003]; [Tamura and Goldenberg 1996]). The prevalence of Zn deficiency in developing countries is similar to that of Fe deficiency, since the same dietary pattern, a reliance on polished rice with minimal dietary diversity, contributes to both ([Black et al. 2008]). In Central, South and South-East Asia and sub-Saharan Africa, stunting due to Zn deficiency affects 40% of preschool children ([Hotz and Brown 2004]), and 82% of pregnant women ([Bhutta and Haider 2009]). Most Latin Americans living in poverty consume a diet rich in cereals and beans, and low in animal products ([Berne Pena et al. 2008]), and therefore record high levels of Zn deficiency ([Amaya et al. 2002]; [WHO 2008]).

Vitamin A deficiency (VAD) is as a public health problem among preschool-aged children in 118 developing countries around the globe ([West 2002]). Vitamin A deficiency occurs when serum retinol is less than 0.7 μmol/l, or less than 20 μg/dl in children below 6 years. The prevalence of VAD among school-aged children (5-15y) in Latin American, South and Southeast Asian countries varies from 6% in Sri Lanka to 36% in El Salvador ([Amaya et al. 2002]; [West 2002]).

Vitamin A plays a major role in phototransduction, and deficiency leads to xerophthalmia followed by complete blindness ([Mason et al. 2005]). Beta-carotene is the precursor of Vitamin A. In a survey of 3000 varieties of rice, only 20 varieties were found to contain -carotene ([Fitzgerald 2007]). The amount of -carotene detected in those 20 was less than 0.2 ppm, and it was found only in the bran layer; no -carotene was detected in the polished grains of any of the varieties ([Fitzgerald 2007]). Therefore people deriving most of their calories from polished rice are at the highest risk of VAD, since other staples, such as certain varieties of wheat, maize and orange-fleshed sweet potatoes contain

-carotene.

Solutions to Malnutrition through Rice

In the context of such widespread malnutrition in the world's major rice consuming countries, agricultural strategists recognised a potential role for rice, if breeding programs could elevate the micronutrient levels in rice ([Bouis and Hunt 1999]). This led to a paradigm shift in breeding programs, as selection for nutritional traits commenced, with the objectives of elevating Fe and Zn, and incorporating -carotene into polished grains. This program was first launched under the umbrella of the Consultative Group for International Agricultural Research (CGIAR) Micronutrient Program, which then underwent reform in 2004 to become Harvestplus, an autonomous organisation that sets the targets and coordinates the multilateral efforts to increase Fe, Zn and -carotene in staple crops, including rice (http://www.harvestplus.org).

The minerals in unpolished rice of 60 popular varieties (Table 1) show that there is less than 10% variability for all minerals except Zn, Mn, and Cu. By exploring wider sources of diversity within the species, significantly more variability was found for the amount of Fe and Zn in unpolished rice ([Gregorio 2002]). However it is not always possible to extrapolate data from unpolished rice to white, polished rice. By replotting data of 70 varieties of unpolished rice and the same varieties polished to a degree of milling of 10% ([Bounphanousay 2007]), it can be seen that the Fe content of all the polished rices clusters around 4 – 5 ppm, despite variation in Fe before removal of the bran layer (Figure 1). By contrast, the Zn content of the unpolished grains shows an association with the Zn content when the grains are polished, and also shows a range in Zn contents of polished rice from 10 – 20 ppm (Figure 1).

Table 1: Amount of selected micro and macro minerals in unpolished grains of 60 popular rices from South-East Asia (Data from Indrasari 2002)

	Ca	P	Mg	K	S	Fe	Zn	Mn	Cu
Amount (ppm)	92.44	3720.59	1488.82	2823.53	1309.12	11.69	23.89	32.51	3.07
Standard deviation	14.13	243.43	103.59	264.07	111.37	1.71	3.96	7.25	1.04

Dipti et al.

Dipti et al. Rice 2012 5:16, doi: 10.1186/1939-8433-5-16

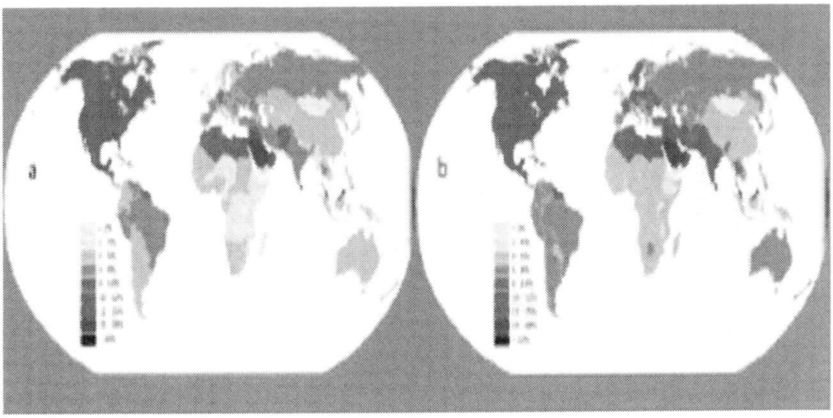

Figure 1: Fe (squares) and Zn (circles) content of unpolished and polished rice of seventy varieties of rice.

In the past ten years, significant international effort has been expended to search the diversity of the species for elevated Fe and Zn in the polished grain ([Hunt et al. 2002]; [Jiang et al. 2007]; [Liang et al. 2007]; [Liang et al. 2008]; [Liu et al. 2004]; [Ma et al. 2005]; [Tucker 2003];[Vasconcelos et al. 2003]; [Wang et al. 2004]; [Welch and Graham 2004]). As suggested by Figure 1, variation for Zn content was discovered, and focussed selection and careful phenotyping has enabled rice improvement programs to release varieties with elevated Zn ([Virk and Barry 2007]). Models using the levels of Zn that can now be achieved in polished rice suggest that the increase in Zn will lead to a significant decrease in the prevalence of Zn deficiency in both adults and children in rural Bangladesh ([Arsenault et al. 2010]).

After searching widely through the diversity of the species, and its wild relatives, rice improvement programs have not found variation in the endosperm for Fe. This suggests that there is either physiological regulation of the Fe that exchanges from the maternal to the filial tissues; or no biological reason, therefore mechanism, for Fe to accumulate in the endosperm. Rice improvement programs have therefore concluded that elevating Fe in the grain can only be achieved by the use of transgenic techniques ([Johnson et al. 2011]).

Transgenic technology successfully elevates Fe in the grain, suggesting no physiological barrier regulating it. In one study, a ferritin gene from soybean was expressed in the endosperm, leading to the accumulation of Fe ([Qu et al. 2005]; [Vasconcelos et al. 2003]). Another study pyramided a ferritin gene from common bean into the grain to increase Fe, with a phytase gene from *Aspergillus* to increase bioavailability of the Fe ([Lucca et al. 2002]). More recently, one study showed a significant increase in Fe content of polished grains by expressing a nicotianamine synthase (*NAS*) gene from barley in the rice endosperm ([Lee et al. 2009]). In another study, the *NAS* genes from rice, *OsNAS1*, *OsNAS2* and *OsNAS3*, that usually express in roots and shoots, were expressed in the endosperm leading to levels of Fe in polished grains that meet and exceed the Harvestplus targets ([Johnson et al. 2011]). Taken together, these studies show that it is possible for iron to enter the endosperm in different forms, and the studies all indicate that a mechanism to import Fe into the endosperm has simply not previously evolved. The ability for different forms of Fe to accumulate provides options to maximise its bioavailability.

The small amount of iron that occurs naturally in the grain and the aleurone layers of the endosperm was unable to reverse Fe-induced anaemia in women ([Haas et al. 2005]). Transgenic rice with the ferritin gene from soybean was able to reverse anaemia in rats with the same efficiency as $FeSO_4$ ([Murray-Kolb et al. 2002]), but the same rice had no effect on iron status of piglets ([Schaffer et al. 2004]). These contradictory findings suggest that bioavailability of ferritin is complex. By contrast, the Fe chelated in nicotianamine was found to be bioavailable to humans ([Zheng et al. 2010]). This suggests an argument for testing the most powerful gene for elevating endosperm Fe, *OsNAS2* ([Johnson et al. 2011]), in other genetic backgrounds, to test (i) if the phenotype is the same in different genetic backgrounds, (ii) for any negative effect on grain yield and grain quality, and (iii) whether the next generation of seedlings, nourished by the endosperm, suffers any effect due to the presence of the iron in the endosperm. Varieties expressing *OsNAS2* in the endosperm possibly offer a solution, other than by reason of bioavailability, to reversing Fe-anaemia. The other transgenic examples use genes from other species to accumulate Fe, whereas those with *OsNAS2* in the endosperm accumulate Fe using a gene that naturally occurs in rice. Transgenic technology was used only to change the expression pattern of that gene. This difference in the degree of transgenic technology, where transgenic techniques were only used to change where the gene expresses in the plant, could ease the road to deregulation and the grains might be acceptable to a wider group of consumers.

A solution to VAD is being tested using transgenic technology to insert the genes necessary to enable the rice grain to accumulate -carotene, creating Golden Rice ([Beyer et al. 2002];[Grusak 2005]). The transgenes have recently been crossed into popular and high-yielding varieties using conventional breeding techniques (http://irri.org/news-events/hot-topics/golden-rice). Recently, a preliminary feeding trial in the US showed that the β-carotene from Golden Rice is efficiently converted to retinol ([Tang et al. 2009]). However, there is very little other information in the public arena about the effect of the transgenes on nutritional, sensory and postharvest quality of the grains of Golden Rice. The high bioavailability, relative to some other sources, was ascribed to the simple starch matrix of the rice ([Tang et al. 2009]). In carrots, β-carotene is located in crystalline chromoplasts where it is less bioavailable than the β-carotene from mangoes, which is held in lipid droplets ([Brackmann et al. 2011]). The location of β-carotene in rice grains is unknown, but its high bioavailability suggests that it is not in chromoplasts. This could also mean that the β-carotene is susceptible to auto-oxidation ([Ramakrishnan and Francis 1979]). After harvest, rice is usually dried to about 12% moisture for storage. In a starch system containing crystalline β-carotene, and stored at room temperature for four weeks at 11% moisture, a 30% loss in β-carotene, due to auto-oxidation was found ([Ramakrishnan and Francis 1979]). Presumably the creators of Golden Rice have achieved a balance between bioavailability and loss due to auto-oxidation, in order to maximize the potential impact on the consequences of

VAD, and have fully characterized any oxidation products that could accumulate in grains. Wider performance testing, including yield trials as well as clinical trials with VAD-deficient target populations in developing countries will be conducted in the near future to verify both bioavailability and potential health benefits of Golden Rice (http://irri.org/news-events/hot-topics/golden-rice). Assuming that all regulatory requirements will be met, IRRI projects that Golden Rice may be ready for release in 2013 (http://www.irri.org).

CHRONIC DISEASES

Chronic Diseases in Rice Consuming Countries

The total number of people with Type II diabetes mellitus (DM) is projected to rise from 171 million in 2000 to 366 million in 2030 ([Wild et al. 2004]). Figure 2 shows the prevalence of Type II DM in each country in 2010, and the projected distribution by 2030 ([Shaw et al. 2010]). Saudi Arabia and North America have the highest prevalence (Figure 2). By 2030, a significant increase in prevalence is seen for Latin America, Africa, and South and Southeast Asia ([Shaw et al. 2010]). One of the major risk factors is that developing countries are changing their eating habits ([Misra et al. 2010]), leading to an obesity epidemic, termed the nutrition transition ([Astrup et al. 2008]; [Hossain et al. 2007]; [Yoon et al. 2006]). In many developing countries, low fat diets are being replaced by fast foods that are high in fat, and this is leading to significant increases in obesity ([James 2008]). The increase in the proportion of the population with obesity is one of the greatest risk factors for diabetes, coronary disease and some cancers ([Kopelman 2000]). Coupled with population growth, rice consuming countries are heading towards a major public health crisis, with significant financial risk at both the household and national levels.

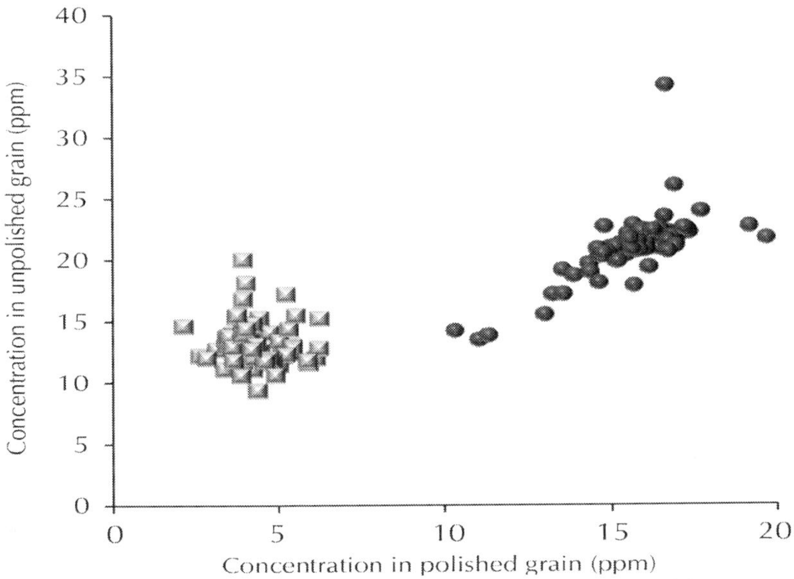

Figure 2: Global distribution of the prevalence of type II diabetes in 2010 (a) and project in 2030 (b). Replotted from the Diabetes Atlas data (www.idf.org/diabelesatlas).

Each year, an estimated 2.5 billion cases of diarrhoea occur among children under five years of age ([Zaidi et al. 2004]). Significantly, the median incidence of diarrhoeal diseases in children under five in developing countries has changed little since the early 1990s ([Jamison et al. 1993]; [Parashar et al. 2003]). More than half of these cases are in Africa and South-East Asia, where bouts of diarrhoea are more likely to result in death or other severe outcomes, such as significant loss of vitamins and minerals ([Udomkesmalee et al. 1990]). By placing these figures and trends in the context of the United Nations Millennium Development Goal 4, which aims to reduce childhood mortality by two-thirds between 1990 and 2015, it becomes clear that many countries in South-East Asia might not meet this target ([You et al. 2010]). In addition, these data provide an illustration of the vicious cycle of malnutrition, where chronic diarrhoea could prevent children from reaping nutritional benefits of rice varieties, developed especially to address malnutrition, with elevated concentrations of vitamins and minerals.

In 2000, more than half the 16.7 million world deaths from cardiovascular diseases were in developing countries, ([WHO 2001]) many of these in South and South-East Asia. Coronary deaths in India are expected to double over the next 20 years ([Ghaffar et al. 2004]), and they reached 2 million in 2010. Cardiovascular diseases were the leading cause of death Malaysia ([Statistics 2010]), which is mainly due to hypercholesterolaemia and hypertension ([Statistics 2006]; [Yunus et al. 2004]). In 1998, the cause of death from cardiovascular disease in Bangladesh was 12.5%, Bhutan 14.8%, India 13.0%, Indonesia 23.2%, Maldives 24.8%, Myanmar 12.7%, Nepal 1.9%, Sri Lanka 20.1%, and Thailand 17.1% ([WHO 2002]).

Cardiovascular disease is a far greater public health problem in developing countries than previously realized, particularly in younger people, according to a report that combined population estimates for five lower to middle income countries with current death rates and workforce data to calculate the effects of cardiovascular disease on society and on the workforce. A conservative estimate showed that at least 21 million years of future productive life were lost each year in the five countries because of cardiovascular disease. Future predictions were even more disturbing, with the number set to rise to 34 million years of life lost by 2020 ([London 2004]).

The cost of managing cardiovascular diseases has been increasing in several countries and regions of the world ([Leal et al. 2006]). One study estimated the burden of cardiovascular diseases in 24 countries of the European Union, and reported the financial burden of cardiovascular diseases was 169 billion per year, with direct healthcare costs accounting for 62% of the cost, followed by costs of informal care, and indirect costs associated with loss of production due to early death, and the loss of productivity due to morbidities ([Leal et al. 2006]).

Potential Solutions to Chronic Diseases from Rice

Whole-grains are the unpolished version of cereal grains, consisting of the germ, bran, and endosperm; while polished grains lack both the germ and bran. The scientific community generally considers whole-grain foods to include those that contain the same amount of germ and bran that would typically be found in the unprocessed grain ([Seal et al. 2006]). By this definition, whole-grain rice, whether consumed intact or pulverised into flour, is a whole-grain food. Consequently, epidemiological studies designed to identify dietary patterns associated with reduced incidence of chronic disease typically classify unpolished, or brown, rice as a whole-grain ([Jacobs et al. 2007]; [Wang et al. 2007]). There is a significant corpus of research describing components in whole-grain rice which have potential for nutritional impact ([Amissah et al. 2003]; [Eggum 1979]; [Goffman et al. 2003]; [Rukmini and Raghuram 1991]; [Shen et al. 2009]; [Storck et al. 2005]).

Whole-grains are hypothesised to contribute positively to human health due to their fibre, minerals (Table 1), vitamins (e.g., vitamins B, D and E), phenolic compounds, phytoestrogens (lignans), and other phytochemicals ([Slavin et al. 1999]). These compounds may influence biological functions individually or synergistically. Whole grain rice contains similar types of compounds to other cereal grains albeit with a few unique types and in unique percentages.

Epidemiological studies suggest consuming whole-grains provides a protective effect against several chronic diseases. Whole-grain rice contains unique types and amounts of some phytochemicals such as the gamma-oryzanol and tocotrienol fractions. Numerous cell culture, animal, and human-based studies have demonstrated the potential health benefits of consuming whole-grain rice and some of its phytochemical fractions. The majority of these studies have focused on the reduction of risk factors for cardiovascular disease ([Cicero and Derosa 2005];[Hallfrisch et al. 2003]), Type II diabetes ([King et al. 2008]; [Topping 2007]; [Zhang et al. 2010]), and several cancers ([Williams and Hord 2005]). By far, the strength of the evidence lies in there being a positive relationship between whole grain rice consumption and reduced risk of disease. Some of the studies supporting these associations, however, have dealt with specific constituents in rice bran and endosperm, and have involved administration of fractions at higher intakes than would be practical from consuming whole grain rice. However, it is possible that these fractions may impact biological targets synergistically and thus exert much higher chemo-preventive efficacy than that found for individual compounds. This possibility may be even more likely when whole grain rice is consumed as part of the daily diet over a long period of time.

The bran components with potential nutritional value include the vitamin E complex of unpolished rice, which is unusually high, ranging from 179–389 mg/kg bran, with an average of 72.5% of the isomers being tocotrienols ([Bergman and Xu 2003]). Other cereals contain much lower average amounts of Vitamin E compared with rice: wheat 23 mg/kg, barley 8 mg/kg, spelt 18.1 mg/kg, and rye 11.9 mg/kg ([Nielsen and Hansen 2008]). The lipid content of rice bran is also high in comparison to other grains. Its primary unsaturated fatty acids are oleic, linoleic and alpha-linoleic, while its primary saturated fatty acids include palmitic and stearic acids. The non-saponifiable fraction of rice bran oil contains tocotrienols, tocopherols, phytosterols, gamma-oryzanol compounds, policosanols, and saponines. Each of these phytochemical fractions consists of several compounds. For example, the gamma-oryzanol fraction is composed of ferulic acid esters of triterpene alcohols. The three primary compounds are cycloartenyl ferulate, 24-methylenecycloartanyl ferulate and campesteryl ferulate (Xu and Godber 1999]), and there are at least seven more compounds in that fraction ([Akihisa et al. 2000]; [Xu and Godber 1999]). Rice bran also contains phenolic compounds which reportedly vary a great deal in quantity and type across different cultivars ([Goffman and Bergman 2002]; [Goffman and Bergman 2004]).

Cardiovascular Disease and Whole-Grain Rice

Studies on the potential health promoting properties of rice on cardiovascular diseases began more than four decades ago ([Nakamura 1966]). These studies report positive effects of whole-grain rice (and several of its fractions) consumption on cardiovascular disease risk factors, such as hypertension and cholesterol, using rodents, rabbits, non-human primates, and humans ([Cicero and Derosa 2005]). The association between whole-grain consumption and protection against heart disease and stroke is considered unequivocal by many, but the exact mechanism is not clear ([Flight and Clifton 2006]).

Hypertension

Hypertension is a significant risk factor for coronary disease. The Dietary Approaches to Stop Hypertension (DASH) diet recommends those with hypertension to increase consumption of whole-grains ([Lochner et al. 2006]). These recommendations are based on the findings of cross-sectional studies examining the correlations between lifestyle and the development of cardiovascular disease. Conclusions from studies examining the specific effect of whole grain consumption on blood pressure, however, have been inconsistent ([Davy et al. 2002]; [Pins et al. 2002]). Many of these studies did not control dietary composition except for whole-grain content and some used whole-grain fractions as opposed to whole-grain foods. By including whole-grain rice as a focus, the design of many of the previous studies evaluating whole-grain consumption and hypertension improved ([Hallfrisch et al. 2003]). Non-hypertensive men with elevated plasma cholesterol levels were fed an American Heart Association Step 1 diet with or without inclusion of unpolished rice/whole wheat, barley, or a combination in a Latin square design. Also controlled were levels of protein, calcium, magnesium, sodium, and potassium in the diets. Systolic, diastolic, and mean arterial blood pressures were reduced in those who consumed soluble fibre from barley or insoluble fibre from unpolished rice and whole wheat, and consumption of the Step 1 diet without the whole-grain component did not have any effect on blood pressure ([Hallfrisch et al. 2003]).

Serum Lipid Levels

A large number of animal, nonhuman primate, and human-based studies provide strong evidence that rice bran and its fractions lower serum cholesterol and triglyceride levels ([Cicero and Derosa 2005]). One study, using 18 humans with moderately-high blood cholesterol levels, were fed 100 g per day rice bran or oat bran for two 3-week periods in a crossover design ([Hegsted et al. 1993]). Prior to each bran phase, a control diet without bran was provided. Total cholesterol levels decreased when rice bran or oat bran was consumed, though neither of the brans had a significant effect on HDL- and VLDL- cholesterol or triglycerides ([Hegsted et al. 1993]). Another human study evaluated the effect of rice bran, oat bran and a rice starch placebo on moderately hypercholesterolemic, non-smoking, non-obese adults during a 6-week, randomised, double-blind trial ([Gerhardt and Gallo 1998]). The 23 males and 21 females were given 84 g of product per day to consume in addition to their regular diet. Significant total cholesterol reduction and improvement in the total cholesterol to HDL-cholesterol ratio in most of these individuals who consumed the bran was reported. Again, there was no significant difference between the effectiveness of the rice and oat bran ([Gerhardt and Gallo 1998]). Both these studies offer rice consumers an option that does not include changing their preferred staple.

Studies using hamsters and nonhuman primates have concluded that the rice bran oil fraction gives rice bran its cholesterol lowering properties ([Kahlon et al. 1991]; [Nicolosi et al. 1991]). Studies using moderately-hypercholesteroleimic healthy humans (n = 26) in a parallel arm design and a randomised crossover design, have compared the effects of a diet including dietary fibre from rice bran or defatted rice bran, and dietary lipids from rice bran oil or another oil blend with a fatty acid composition similar to rice ([Most et al. 2005]). The study showed that defatted rice bran did not lower lipid levels, and consumption of the diet containing rice bran oil compared to the control diet resulted in lower total cholesterol levels ([Most et al. 2005]), suggesting that the oil in rice bran contains unique compound/s for lowering cholesterol. The fractions of primary focus for this capacity have been the tocotrienols, sterols, gamma-oryzanol and policosanols.

Less than 1% of all research published regarding the Vitamin E complex has focused on tocotrienols. Reviews indicate that most of the research on tocotrienols has focused on this fraction from palm oil or individual isomers, while only a limited focus has been placed on this fraction in rice bran ([Packer et al. 2001]; [Rasool and Wong 2007]; [Sen et al. 2007]). Research with cell cultures has shown that tocotrienols together and as individual isomers influence cholesterol synthesis by regulating the expression of 3-hydroxy-3-methylglutaryl-coenzyme A reductase, the rate-limiting enzyme in the cholesterol synthesis pathway ([Parker et al. 1993]). Interestingly, -tocopherol has shown an opposite effect in hypercholesterolemic human subjects ([Qureshi et al. 2002]).

Conflicting conclusions have been reported from clinical trials that examined the effects of rice and palm tocotrienol rich fractions on cholesterol ([Packer et al. 2001]; [Rasool and Wong 2007];[Sen et al. 2007]). The studies that reported an inhibitory effect of tocotrienols on total cholesterol levels used preparations with less than 20% tocopherols. The reverse was reported for studies that used preparations with a greater percentage of tocopherols. Rice bran reportedly contains approximately 25% tocopherols and 75% tocotrienols ([Bergman and Xu 2003]), so this could be one of the ways that rice bran lowers cholesterol. Potentially confounding aspects of the design of human trials have been that the relative amounts of the four tocotrienol isomers varied between these studies. Several of these studies, also, did not control the amount of dietary lipids and alcohol consumed by the subjects; both have been reported to modulate the effects of tocotrienols ([Qureshi et al. 1997]).

Policosanols are a mixture of primary long-chained alcohols. Sugarcane policosanols reduced plasma LDL cholesterol in several clinical trials of varying duration and at efficacious doses, ranging from 2 to 40 mg per day ([Chen et al. 2005]; [Varady et al. 2003]). Whole-grain rice contains policosanols but its individual compounds are found in different ratios compared to the similar fraction in sugarcane. Rice policosanols (10 mg per day), when fed to hypercholesterolemic men and women in a randomised, double-blind, crossover, placebo-controlled trial (n = 70), reportedly lowered the subjects plasma total cholesterol, and increased levels of Apolipoprotein A-1, the major protein componenet of HDL ([Reiner et al. 2005]). However, a lack of cholesterol-lowering efficacy of sugarcane policosanols was reported from a study with a similar design to those reported above ([Berthold et al. 2006]). Thus, the association between policosanol consumption and plasma cholesterol reduction is unclear. To clarify this situation, future research will need to take into consideration that the specific compounds and amounts of each type in the policosanol fraction vary both between and within crops.

Numerous studies with rodents have reported that the gamma-oryzanol fraction from rice bran is able to lower serum cholesterol levels in animals fed different model hypercholesterolemic diets ([Cicero and Derosa 2005]). The mechanism of action appears to include increased faecal excretion of cholesterol and its metabolites ([Wilson et al. 2007]). These authors also reported that ferulic acid from rice bran showed anti-atherogenic properties, but through a different mechanism. The serum cholesterol lowering properties of gamma-oryzanol have not been confirmed in humans.

While it seems clear that there is an association between the consumption of whole grain rice and a lowering of cardiovascular risk factors, the mechanisms leading to this could be due to multiple compounds. In order for rice improvement programs to make use of these associations, there is a need for investment targeted specifically towards identifying the mechanism of risk reduction, understanding variability within rice for managing cardiovascular risk factors, and then using that information to develop phenotyping tools so that selection for heart-healthy varieties of rice is possible in the future.

Cancers and Whole Grain Rice

A meta-analysis of prospective epidemiologic studies suggests that consumption of whole-grain products is inversely associated with the development of several forms of cancer ([Williams and Hord 2005]). However, only a limited number of human, animal and cell culture-based studies that specifically evaluate the association of whole grain rice, or it's fractions, with cancer risks have been done, and many studies cannot be done in humans due to limitations in analytical capability. The studies indicate that whole grain rice contributes to mitigating cancers, tumour growth and proliferation by two different mechanisms. There is an increasing body of evidence that resistant starch (RS) in the whole grain provides one mechanism, and the second is through bioactive compounds that could be present in the bran layer of the rice. However, neither mechanism is well understood, but the results described below suggest that both mechanisms are worthy of pursuing further.

Colon cancers arise from benign neoplasms and evolve into adeno-carcinomas through an histological sequence beginning with either adenomas or hyperplastic polyps. A link between dietary fibre and reduced risk of colon cancer was first proposed several decades ago. That association remains controversial; however, confidence in a link between whole grain consumption and reduced risk of colorectal cancers is growing ([Annison and Topping 1994];[Bird et al. 2000]; [Park et al. 2005]; [Schatzkin et al. 2007]; [Topping et al. 2003]; [Topping 2007]; [Topping and Clifton 2001]). Several mechanisms have been proposed to explain this possible association. Cereal bran may have a protective effect against colorectal cancer by altering the colonic bacteria profile via addition of fermentable carbohydrates, such as resistant starch (RS).

Recent evidence suggests that RS results in an increased production of short chain fatty acids (SCFA) such as butyrate in the colon ([Topping 2007]). These are considered to play a number of roles in bowel health, including recovery from chronic diarrhoea, lowering absorption of potential carcinogens, and repair of damaged DNA ([Topping 2007]). Studies using pigs found faecal SCFA levels higher when the pigs were fed whole grain rice relative to feeding with milled rice and bran. This was reportedly due to greater RS in the whole grain rice wherein the bran layer protected the starch from digestion, enabling the starch to ferment in the large intestine where it produced SCFA ([Bird et al. 2000]). Consistent with this, a systematic study of RS in a diverse set of rices showed that RS in unpolished cooked rice was 30% higher than in the polished cooked rice of the same variety ([Williams et al. 2005]), probably due to the bran layer preventing digestive enzymes from accessing the starch. RS can also be manipulated in polished grain. The starch branching enzyme IIb gene (*SBEIIb*) has a large effect on RS, and varieties that carry a mutation in this gene show elevated RS ([Butardo et al. 2011]), but when expression of that gene is completely silenced with specific transgenic techniques, the RS is significantly increased ([Butardo et al. 2011]).

Familial adenomatous polyposis is a hereditary condition that predisposes people to colon cancer. Apc(Min) mice carry a mutation in the same gene that causes familial adenomatous polyposis in humans. These mice develop large numbers of intestinal tumors at an early age and are thus used as a model for evaluating chemo-preventive interventions for humans with intestinal polyps. When rice bran was included in the diet in a cross-over experiment, the Apc(Min) mice showed a decrease of 51% in the number of intestinal adenomas ([Verschoyle et al. 2007]). The mechanism is unclear, but is likely to be due to compounds present in the bran.

Isoprenoids are known to possess potent anti-cancer activity ([Sen et al. 2007]), and whole grain rice contains several different types of these compounds. For example, γ-tocotrienol, the tocotrienol that rice has in the largest quantity, has been documented using tissue culture techniques to be the most potent anti-cancer Vitamin E isoform of all the isomers that occur in nature ([Sen et al. 2007]). Numerous studies documenting this observation have shown that tocotrienols and γ-tocotrienol in particular, target Nuclear Factor-κB (a transcription factor) which reduces inflammation and thus mediates the impact of carcinogens ([Ahn et al. 2007];[Nesaretnam and Meganathan 2011]). Work using mice not only supports the results found using cell lines, but when extrapolated to humans, indicates that an efficacious dose of tocotrienols could be consumed from the diet ([He et al. 1997]). Other isoprenoids in whole grain rice have been reported to interfere with the colony-forming ability of breast and colon cancer cells ([Hudson et al. 2000]). Eight phenolic compounds, protocatechuic acid ([Hudson et al. 2000]), ρ-coumaric acid ([Zhou et al. 2004]), caffeic acid ([Hudson et al. 2000]), ferulic acid ([Tian et al. 2004]), sinapic acid ([Hudson et al. 2000]), vanillic acid ([Zhou et al. 2004]), methoxycinnamic acid and tricin ([Hudson et al. 2000]), were identified in the extracts studied. Of these compounds, the flavonoid tricin has received the greatest research attention ([Zhou and Ibrahim 2010]) which has likely been due to its greater activity in interfering with cancer cells in tissue culture experiments at levels lower than the other phenolics studied, making it an easier compound to study than others.

During tumour metastasis, a critical early step is cell invasion of the basement membrane - a dense meshwork of collagen, glycoproteins, and proteoglycans which, under normal circumstances, prevents cells from moving away from their sites of origin. Cancer cells, however, secrete several different types of enzymes that digest the proteins in the basement membrane ([Fidler 2003]). When the membrane has been sufficiently weakened, the tumour is able to push through the membrane ([Liotta and Stetler-Stevenson 1991]), which allows cells to invade surrounding tissue. Another isoprenoid fraction from whole grain rice, specifically the anthocyanins cyanidin 3-glucoside and peonidin 3-glucoside were isolated and shown to inhibit the mobility and invasion ability of human hepatocellular carcinoma (SKHep-1) cells ([Chen et al. 2006]). This effect was associated with a reduced expression of several proteinases. In addition, these compounds were fed to mice after subcutaneous inoculation with SKHep-1 cells. Small solid tumours were observed following cell inoculation and a 1.9-fold reduction in tumour volume and a 1.7-fold reduction in tumour weight were reported after feeding the mice the anthocyanin fraction from whole grain rice ([Chen et al. 2006]). The anthocyanin, antioxidant and phenolic content ranges enormously in rice, and several studies associate it positively with coloured pericarps ([Goffman and Bergman 2004]; [Han et al. 2004]; [Oki et al. 2004]; [Shen et al. 2009]). Figure 3shows the diversity in the colour of the rice pericarps.

Figure 3: Diversity in the colour of the bran of unpolished rice.

The studies described above all indicate that whole grain rice plays a significant role in preventing cardiovascular risk factors, tumour growth and tumour proliferation, through both the mechanical protection of starch by the bran with the subsequent benefits of SCFA, and the range of bioactive compounds in bran. These data call for stronger links between rice researchers and the medical fraternity, significant investment in research to detect and identify the important grain constituents and to quantify the nutritional impact in humans, and techniques to deliver to rice improvement programs to enable selection on bran quality. The bran content of rice ranges from 5–8% ([Bergman and Chen 2007]), which could be an immediately selectable trait for rice varieties targeted to markets that consume either whole-grain rice, or different forms of it, such as pre-germinated rice, which is also consumed as whole-grain rice.

Type II Diabetes Mellitus and Rice

Diabetes is a chronic evolving disease associated with a variety of micro- and macro-vascular complications. Although pharmacological therapies are effective, the diabetes prevention trials in Finland and the U.S. remind us that nutrition and lifestyle approaches can be more effective in delaying onset of the disease. In fact, these nutrition and lifestyle approaches to diabetes prevention and treatment should be given at least as much attention as drug therapies. The use of low glycaemic index (GI) foods should be considered as one of a number of tools available to manage, or prevent the onset of, Type II diabetes.

The GI of a food quantifies the rate of release of glucose into the blood in response to the carbohydrates consumed. The glycaemic load (GL) is the mathematical product of the GI and the amount of carbohydrate in the food. A recent meta-analysis of 37 prospective observational studies concluded that GI and GL are both strongly associated with Type II diabetes and its chronic diseases ([Barclay et al. 2008]). Another recent study confirms this ([Halton et al. 2008]), and a prospective cohort study and systematic review of six other cohort studies concluded that two-servings-per-day of a whole grain was associated with a 21% decrease in risk of type II diabetes ([Munter et al. 2007]). These studies consisted of 286,125 participants and 10,944 of these had Type II diabetes. A feeding trial of patients with Type II diabetes found that those on the diet of lower GI had a significant improvement in blood sugar status compared with those on the higher GI diet ([Nisak et al. 2010]), strongly suggesting that choice of carbohydrate will aid in the management of blood sugar status.

Rice has generally been considered to be a food of high GI ([Brand-Miller et al. 1992]). Several studies link the consumption of rice with increased risk of developing Type II diabetes ([Nanri et al. 2010]; [Villegas et al. 2007]). Those studies were carried out in Japan and Shanghai respectively, and in both those places, the rices consumed are likely to be of high GI because they are low amylose varieties ([Sato et al. 2004]; [Fitzgerald et al. 2011]). Both [Nanri et al.] (2010) and [Villegas et al.] (2007) report an association between diabetes and rice consumption, and the association was stronger in people with low physical activity coupled with high intake of rice. Another study based in Australia shows that the risk of developing Type II diabetes was highest in people who consume high levels of white bread together with low physical activity ([Hodge et al. 2004]). The GI of white bread is high ([Foster-Powell et al. 2002]). Together, the three studies indicate that consumption of carbohydrate of high GI is likely to increase risk of Type II diabetes, especially when coupled with low physical activity, irrespective of the grain delivering the carbohydrate. In areas where rice is the staple, correlations between consumption of high GI rice and Type II diabetes will be found, however where wheat is the staple, the correlations with Type II diabetes will be found for those who consume white bread. Furthermore, when the diversity of rice and rice products is explored, a significant range in GI has been demonstrated. For a set of 260 varieties of polished rice, GI ranged from 52 – 92 ([Fitzgerald et al. 2011]). Rice products, such as parboiled rice and rice vermicelli, have also been shown to give a GI of at least 10 units below the GI of the white rice of the same variety ([Ranawana et al. 2009]; [Sato et al. 2010]). These studies suggest that particular varieties of rice and rice products can be chosen to manage Type II diabetes, or possibly prevent the onset of the condition.

It is currently unclear whether whole-grain rice (unpolished) offers any advantage over polished rice for the management of blood glucose. A prospective study in the US reported that consumption of polished rice led to a greater likelihood of developing Type II diabetes than consumption of whole-grain rice ([Sun et al. 2010]). However, [Sun et al. (2010]) compared the average GI of brown and white rice published by [Foster-Powell et al. (2002]), but only three varieties were tested as both brown and white rice. Comparing the GI values of the brown and white rice of the three varieties, Doongara, Amaroo and Pelde, showed no difference in GI for the brown and white forms of Doongara and Amaroo ([Foster-Powell et al. 2002]). Furthermore, other differences between the levels of physical activity and other dietary choices of the brown and white rice consumers studied by [Sun et al. (2010]) could easily contribute to the difference in the incidence of Type II diabetes. Two earlier studies that compared the GI of different varieties of whole-grain (brown) rice with the polished rice of the same variety gave conflicting results. The first study measured GI in eight healthy volunteers who consumed whole-grain and polished rice in a randomised experiment, and no difference in blood glucose response was found between the pairs of whole-grain and polished for the three varieties ([Brand-Miller et al. 1992]). The second study found significantly lower blood glucose responses when participants consumed the whole-grain rice compared to the polished rice of the same variety ([Panlasigui and Thompson 2006]). These differences could be due to varietal differences, differences in cooking, differences in measuring GI, or other factors of diet and lifestyle. It is therefore not yet possible to conclude if the consumption of whole-grain rice gives any advantage over polished rice for the management of blood glucose status.

The Nurses' Health Study and the Harvard School of Public Health indicate that higher intakes of Mg may reduce the risk of developing Type II diabetes ([Lopez-Ridaura et al. 2004]). Research has shown that low levels of Mg may impair insulin sensitivity or function. Consuming adequate levels of Mg may help insulin function properly in the body, which may assist in preventing Type II diabetes. Table 1 shows a significant amount of Mg is found in the whole-grain, but polishing to white rice removes 86% of it ([Hansen et al. In Press]).

The diversity of compounds found in the bran, metabolites and minerals in the grain, structures of starch, non-starch polysaccharides, and different cooking and processing methods are all likely to have some impact on the digestibility of rice. Given that rice consuming countries face a grand challenge with skyrocketing rates of Type II diabetes, it is essential that we understand a lot more about the digestibility of rice and the importance and roles of other bioactive compounds from rice and bran. Investment is needed to enable science to identify and validate important compounds, and deliver this knowledge, as phenotyping tools, to rice improvement programs.

Consumption of Whole Grain Rice

Unpolished rice is not widely consumed world-wide, and is most likely to be consumed in Western countries by health-conscious consumers. It can be found in the markets in most Asian countries, but the belief is that the unpolished rice is for the elderly, and anecdotally it is said to provide nutritional and metabolic benefits. However, there are two methods of processing rice that are likely to provide some of the nutritional value of whole-grain to consumers. The first is parboiling of the paddy before it is polished to white rice. The second is the relatively new technique of pregerminating the brown rice grain prior to cooking the brown rice, which alters the biochemical profile of the bran and improves the cooking and sensory properties of brown rice.

Parboiling involves three basic processes prior to dehulling and polishing: soaking (or steeping), steaming/boiling, and drying ([Chukwu 1999]). Parboiling is practiced in India, Pakistan, Sri Lanka, Bangladesh, West Africa, the Americas and Europe, and the methods used are all slightly different. After parboiling, the rice is polished before it is consumed, and the polished parboiled rice is considered to be of superior nutritional value compared to polished rice that has not been parboiled ([Amato et al. 2002]; [Pedersen et al. 1989]). The nutritional advantage of parboiled polished rice is thought to be due to the leaching of minerals and water-soluble vitamins from the bran layers into the endosperm during the parboiling process ([Amato et al. 2002]; [Juliano 1985]; [Nunes et al. 1991]). However, not all migrating minerals are recovered in the endosperm. Comparisons between the mineral content of brown (not parboiled) and parboiled brown rice of the same variety show slight variations for the concentration of Fe, Zn and Ca, suggesting that a proportion of these minerals was lost during the parboiling process ([Heinemann et al. 2005]). Two studies have also shown a slight decrease in carotenoid content of parboiled rice (0.003 mg to 0.001 mg/100 g) ([Gariboldi 1973]; [Otegbayo et al. 2001]), again, presumably due to leaching. Parboiling leads to the migration of thiamine, riboflavin and bran pigments, such as bioactive anthocyanins, into the grain ([Lamberts et al. 2006]; [Manful et al. 2007]). However, there is no systematic study quantifying (i) leaching from the bran into the endosperm, or (ii) the nutritional value of polished parboiled rice due to different parboiling treatments, and no information can be found about the leaching patterns of the classes of bioactive compounds discussed above.

Pregermination is another processing method that could enable consumers to avail of the nutritional value of bran. In this process, the brown rice is soaked at 37°C for 24 h to initiate the germination process ([Finney 1983]; [Sakamoto et al. 2007]). The rice is then dried, and vacuum-sealed, and it is not polished prior to consumption. Pregermination leads to extensive biochemical changes in the pregerminated rice compared with the ungerminated whole-grain ([Kayahara and Kikuichi 2000]; [Sakamoto et al. 2007]). These changes lead to significant increases in bioactive compounds such as γ-aminobutyric acid (GABA), dietary fibre, inositols, ferulic acid, phytic acid, tocotrienols, Mg, K, Zn, γ-oryzanol, and prolylendopeptidase inhibitor ([Kayahara and Kikuichi 2000]). Total phenolics increased to a maximum level after a germination time of 24 hours while GABA, which was not present before germination, developed in significant quantities upon pre-germination ([Hirunpong and Tungjaroenchai 2008]). Novel acylated steryl glucosides appeared at the same level of bioactivity as found in soybean after pre-germination ([Usuki et al. 2007]). In countries of South-East and North Asia, pregerminated rice is marketed with the nutritional label of GABA rice ([Panchan and Naivikul 2009]).

Germination provides further nutritional value by reducing anti-nutritional factors in whole-grain rice, such as phytate ([Liang et al. 2008]; [Shallan et al. 2010]). Phytate inhibits the bioavailability of minerals by forming stable, indigestible complexes ([Ma et al. 2005]; [Welch and Graham 2004]). The process of germination activates the production of phytase, which catalyses the hydrolysis of phytate ([Sung et al. 2005]). Moreover, the sensory properties of pregerminated rice are considered to be superior to those of brown rice ([Fujino and Kuwata 2004]), and this, together with the nutritional enhancement, could increase the acceptability of pregerminated rice amongst consumers of polished rice.

Exploring genetic variability for the response to pregermination could lead to a wealth of new information about the nutritional potential of rice. Partnerships with the medical profession could enable nutritional impact to be quantified, which could then flow through to rice improvement programs, to improve the nutritional value of rice by maximising the potential of the raw material, and of the process.

New Technologies for Selecting Tools for Nutritional Quality

With the importance and relevance of nutritional components being clear, there is an increasing demand for more detailed information on the molecular mechanisms behind the biochemical content of the rice grain, both before harvest and after polishing. New high throughput phenotyping platforms such as metabolomics, giving unprecedented insights into grain composition, and HTP genotyping platforms such as Next Generation Sequencing, RNAseq and SNP arrays, which reveal the genetics behind varietal differences, have huge potential.

Metabolomics as a New Phenotyping Tool for Nutritional Compounds

Metabolomics approaches for plants have now been around for little more than 10 years ([Fiehn et al. 2000]), and give us the capacity to study the biochemical composition of plant materials in an untargeted manner ([Hall 2006]; [Hall 2011]). The biochemical profiles obtained, using a range of now standard platforms, cover most of the main groups of nutritionally-relevant small molecules. Large (polymer) molecules are not included, which makes metabolomics an excellent complement to other approaches targeting key nutrient groups, such as starch and proteins ([Fernie and Schauer 2009]; [Hall 2006]). Our knowledge of the plant metabolome is still limited, as was demonstrated recently using rice as an example to show, that even after a detailed (literature) analysis of all available information on this crop's metabolites, we must still only have visualised just a small fraction ([Kind et al. 2009]). Metabolomics is helping to increase this coverage, but much work is still needed if we are to link e.g. key nutrient traits with mixtures of known metabolites ([Fernie and Keurentjes 2011]; [Stewart et al. 2011]).

For rice, only a few true metabolomics studies have been published to date (see ([Hall et al. 2008]; [Oikawa et al. 2008]; [Tarpley and Roessner 2007])). Nevertheless, the potential of the technology has clearly been recognised, particularly in areas such as stress tolerance ([Ahuja et al. 2010]; [Tarpley and Roessner 2007]), grain development, grain quality, and nutritional value ([Fitzgerald et al. 2009]; [Yamakawa and Hakata 2010]). While early rice metabolomics was often performed on leaf material (Capillary Electrophoresis MS, ([Sato et al. 2008]; [Sato et al. 2004]), later, additional approaches have been used to analyse rice grains (GC-MS on transgenic rice, ([Zhou et al. 2008]); 1D and 2D GC-MS on brown rice, ([Kusano et al. 2011]); HSSE/GC-MS on rice grains, ([Grimm et al. 2011])). Only one metabolomics study has so far been published on rice bran, but specifically in the context of its use as a suitable starting material for fermentation to produce health-promoting phytochemicals ([Ryan et al. 2011]). Methods for the untargeted analysis of rice grain volatile compounds, which are important in determining key phenotypic traits in fragrant rices such as Basmati and Jasmine types, have also recently been published ([Boualaphanh et al. 2011a]; [Calingacion et al. 2011]; [Verhoeven et al. 2011]). GC-MS results have revealed that rice fragrance (measured from extracts or directly from head space) is created from a rich mix of a wide range of chemically-diverse compounds ([Bryant and McClung 2011];[Calingacion et al. 2011]; [Champagne et al. 2008]). Many of these can be associated with positive fragrance traits, but others can be linked to off-flavour/negative traits. Advanced knowledge of both is essential when we wish to design breeding programmes with increased capability to select for nutritional quality. Using LC-MS, ([Heuberger et al. 2010]) also revealed a vast richness in the non-volatile metabolites, many of which could be associated with nutritional value or potential health benefits.

In the most extensive integrated approach to date, [Calingacion et al., (2011]) have used five complementary metabolomics platforms (NMR, LC-MS, GC-MS, GC-TOF-MS and ICP-MS) to gain a broad insight into minerals and metabolites in polished rice grains. Three contrasting genotypes were used in a proof of concept experiment, and it was demonstrated that all platforms could readily distinguish each genotype. This indicates that irrespective of analytical platform or group of metabolites studied, chemical diversity is sufficient to enable genotype-specific profiles to be identified. This individuality in diversity is presumably also the fundamental basis of the high discriminatory potential of the human palette to distinguish rice varieties after cooking, even when the grains come from genetically closely related varieties ([Calingacion et al. 2011];[Champagne et al. 2010]). Of course, metabolic complexity is, to a large extent, the reflection of complexity resulting from genetic polymorphism. Our growing potential to exploit the sequence of the rice genome, in combination with e.g. full genome transcriptomics, comprehensive SNP genotyping and genome-wide association studies using multi-platform metabolomics studies ([Calingacion et al. 2011]), is predicted to enable us to make unprecedented steps in our quest to understand better the molecular basis of rice metabolite profiles, and design tailor-made nutrition-directed breeding strategies for specific rice varieties, whole-grain and polished, to meet future nutritional needs.

New Genotyping Tools for Defining the Genetics of Nutritional Quality

Over the last ten years, genotyping tools have advanced immeasurably. Genome-wide genotyping has become almost routine, and now the current advances are increasing the resolution and application of the technology to rice exponentially ([McCouch et al. 2002]). Now, single nucleotide polymorphisms (SNPs) are replacing most genotyping techniques, and with new SNP techniques, DNA from a single variety can be screened at many loci in a single pass ([Boualaphanh et al. 2011b]; [McCouch et al. 2010]).

Routine genotyping at 384 SNP loci is currently possible for rice, for both *indica* and *japonica*subspecies ([Boualaphanh et al. 2011b]; [Wright et al. 2010]). However, just as genotyping at 384 loci has become routine, the number of loci on newly developed chips has risen to 44000, and is soon to reach 1, 000,000 loci ([Tung et al. 2011]). As this rich collection of SNPs are discovered, the genotyping can be used to develop mapping populations rapidly ([Boualaphanh et al. 2011b]), or used to associate with phenotype data ([Calingacion et al. 2011]), or a particular set of SNPs can be selected and used for specific genotyping in a breeding program ([Chen et al. 2011]).

Once phenotyping tools can identify compounds and structures that offer nutritional benefit, the new genotyping tools offer the opportunity for rapid discovery of the genes responsible for the phenotype. Such genetic knowledge can then be delivered to breeding programs to enable genetic selection for compounds, structures and traits that offer nutritional benefit.

Consumer Acceptance of Nutritionally Enhanced Rice

Consumers have eaten particular varieties for many generations in the belief that they give nutritional value. For example, rice in India with red bran is marketed and prescribed by Ayurvedic practioners for its health benefits such as controlling hypertension and diabetes ([Ahuja et al. 2008]). In Laos, specific varieties are consumed by post-partum women ([Bounphanousay 2007]). Belief in these effects on health has been upheld for many generations, but the mechanisms through which they act are not clear. Any rice that is sold unpolished in the U.S. can legally be marketed with the following FDA-approved health claim: "Diets rich in whole-grain foods and other plant foods and low in total fat, saturated fat, and cholesterol may reduce the risk of heart disease and some cancers" ([FDA 1999]). However, labels are not specific, mechanisms are not well understood, impact on health is not quantified, and consumption of whole-grain rice is very low in developing countries. The purpose of the present review is to create awareness of the nutritional potential of rice; identify current progress with developing varieties for specific issues of malnutrition and chronic disease, through the understanding the constituents of the rice grain, and bringing new science to bear; and identifying new opportunities to tap into the diversity of rice and identify or develop additional nutritionally enhanced varieties. However, there are a number of issues to consider when marketing such varieties to policy makers, farmers and most importantly, consumers.

Some countries have very strict food labeling regulations, such as in Japan, while others have relatively lax laws, as in China ([Hawkes 2004]). In regions of the world with strict regulations, marketing rice as possessing a certain level of a nutrient will require that the levels are documented through laboratory testing. Labeling rice with a health claim will be particularly challenging since countries have diverse requirements for making such claims, and consumers believe claims to varying degrees ([Shimizu 2011]). Furthermore, in most regions of the world, cultivars with an enhanced nutrition profiles, such as those with elevated Zn, will need to be segregated from those without the trait so the impact is not diluted. Segregation of particular cultivars is known as identity preservation (IP) across the crop world. IP requires cultivars to be harvested, stored, and marketed separately ([Sundstrom et al. 2002]). These systems specify that particular standards be upheld, records kept, and testing be performed. In other situations IP is used to enable consumers to choose products without a particular trait, as in the case of genetically modified (GM) crops. The European Union, Japan, Australia and other regions of the world have laws requiring the labeling of GM foods ([Huffman 2004]). Thus, IP will likely be required for a nutritionally enhanced rice to be sold whether it is created using traditional plant breeding techniques, mutation breeding, or genetic engineering.

Bringing a nutritionally enhanced rice to a consumer or a patient has its own set of obstacles. The first of these is that it is necessary to meet some of the nutritional challenges using the tools of transgenic technology. Golden rice was developed with genes from another species (maize) ([Grusak 2005]), but the high Fe rice that best meets nutritional targets was developed using a transgenic technique to enable a gene found in all rice varieties, that usually expresses in the roots and shoots, to express in the grain as well ([Johnson et al. 2011]). These are different degrees of genetic modification, but will that influence consumer choice? The development of transgenic food products has been highly controversial, in spite of the fact that there was no other way to create grains of rice with nutritionally useful levels of iron or rice grains containing -carotene. Generally consumer skepticism is focused around questions of unknown environmental and health consequences of growing or consuming the transgenic products ([Curtis et al. 2004]). The two transgenic products described here have been developed with the specific humanitarian objective of improving the health of the poorest people. However, consumers and policy makers have every right to ask questions, and should do so. As more trials are undertaken in different locations and in target populations, science will begin to provide the answers to the critical questions, which then will enable policy makers and consumers to make informed decisions.

A second obstacle to consumer acceptance concerns the capacity of science to quantify the nutritional impact of the nutritionally valuable rices, especially with regard to chronic diseases, and to communicate that effectively to the medical community. At present, studies investigating the impact of nutritionally enhanced rice in preventing chronic disease, or developing therapeutic strategies for controlling these, are lacking, and specific dietary recommendations are not focal points in the management of chronic diseases by the medical community.

Presently, the medical community primarily practices evidence-based medicine, which aims to apply the best available evidence gained from the scientific method to clinical decision-making. Quantifiable, evidence-based data based on extensive research would be invaluable for encouraging more medical practitioners to combine dietary recommendations with medicinal treatment as their primary strategy in managing patients with chronic diseases. However current limitations to this are the identification of grain constituents with health benefits, and the existence of rapid and accurate methods that enable data to be gathered to quantify medical impact. Techniques for testing nutritional claims from rice research would provide the conduit for collaboration between the medical community and rice scientists to enable rice varieties to be developed to provide solutions to chronic diseases.

CONCLUSIONS

Prioritising investment into identifying the biological causes underlying associations between rice consumption and decreased risk factors for chronic diseases would have significant and long-term impact on global nutritional challenges that have their greatest effect in developing countries. Technologies are advancing at rates rapid enough to make scientific progress in these areas, and doing so would lead to both nutritional and economic benefit in every rice-consuming country. The outputs from such research programs must be integrated with other efforts aiming to deliver climate-ready varieties that resist the challenges of the changing environment, and varieties that are acceptable to consumers. Rice that does not meet the needs of consumers in terms of physical and sensory properties will not be successful in the marketplace, but once research programs are resourced to understand the linkages between chemicals in rice and human health, and are able to deliver solid nutritional information to the medical community, consumers might be able to be persuaded to increase their consumption of wholegrain or pregerminated rice.

AUTHORS' CONTRIBUTIONS

DSS, CB and MF made the major contribution to writing the manuscript. All other authors made equal contributions, edited and approved the manuscript.

ACKNOWLEDGEMENTS

The authors thank the International Diabetes Foundation for making their data publicly available, and Dr Andrew Nelson from IRRI for mapping the data for Figure 2. We also thank Anna Genil from IRRI for assistance.

REFERENCES

1. Ahn KS, Sethi G, Krishnan K, Aggarwal B (2007) Gamma-tocotrienol inhibits nuclear factor-kappaB signaling pathway through inhibition of receptor-interacting protein and TAK1 leading to suppression of antiapoptotic gene products and potentiation of apoptosis. J Biol Chem 282:809-820
2. Ahuja U, Ahuja SC, Thakrar R, Singh RK (2008) Rice – A Nutraceutical. Asian Agri-History 12:93-108
3. Ahuja I, de Vos CHR, Bones A, Hall RD (2010) Plant molecular stress programs face climate change. Trends Plant Sci 15:664-674
4. Akihisa T, Yasukawa K, Yamaura M, Ukiya M, Kimura Y, Shimizu N, Arai K (2000) Triterpene alcohol and sterol ferulates from rice bran and their anti-inflammatory effects. J Agric Food Chem 48:2313
5. Amato GW, Carvalho JLV, Silveira (2002) parboiled rice fs: clean, noble product. Rio Grande do Sul Porto Alegre.
6. Amaya D, Viloria H, Ortega P, Gómez G, Urrieta J, Lobo P, Estévez J (2002) Vitamin A deficiency and the anthropometric nutritional status of urban and rural marginalized children in the state of Zulia, Venezuela. Investigation Clinic 43:89-105
7. Amissah JGN, Ellis WO, Oduro I, Manful JT (2003) Nutrient composition of bran from new rice varieties under study in Ghana. Food Control 14:21-24
8. Annison G, David L, Topping (1994) Nutritional role of resistant starch: chemical stucture vs. physiological function. Annu Rev Nutr 14:297-320
9. Arsenault JE, Yakes EA, Hossain MB, Munirul Islam M, Ahmed T, Hotz C, Lewis B, Rahman AS, Jamil KM, Brown KH (2010) The current high prevalence of dietary zinc inadequacy among children and women of rural Bangladesh could be substantially ameliorated by zinc biofortication of rice. J Nutr 140:1683-1690
10. Astrup A, Dyerberg J, Selleck M, Stender S (2008) Nutrition transition and its relationship to the development of obesity and related chronic diseases. Obes Rev 9:48-52

11. Barclay AW, Petocz P, McMillan-Price J, Flood VM, Prvan T, Mitchell P, Brand-Miller JC (2008) Glycemic index, glycemic load, and chronic disease risk–a meta-analysis of observational studies. Am J Clin Nutr 87:627-637
12. Bergman C, Chen M (2007) Not all rice bran is created equal. beneath the hull: exploiting the health-beneficial properties of the rice grain. Rice Utilization Workshop, New Orleans, LA.
13. Bergman C, Xu Z (2003) Genotype and environment effects on tocopherol, tocotrienol, and gamma-oryzanol contents of Southern US rice. Cereal Chemistry 80:446-449
14. Berthold HK, Unverdorben S, Degenhardt R, Bulitta M (2006) GouniBerthold. 2006. Effect of policosanol on lipid levels among patients with hypercholesterolemia or combined hyperlipidemia: a randomized controlled.trial. J Am Med Assoc 295(19):2262-2269
15. Beyer P, Al-Babili S, Ye X, Lucca P, Schaub P, Welsch R, Potrykus I (2002) Golden Rice: Introducing the -carotene biosynthesis pathway into rice endosperm by genetic engineering to defeat Vitamin A deficiency. J Nutr 132:506S-510S
16. Bhutta ZA, Haider BA (2009) Prenatal micronutrient supplementation: Are we there yet? Can Med Assoc J 180:1188-1189
17. Bird AR, Hayakawa T, Marsono Y, Gooden JM, Record IR, Correll RL, Topping DL (2000) Coarse brown rice increases fecal and large bowel short-chain fatty acids and starch but lowers calcium in the large bowel of pigs. J Nutr 130:1780-1787
18. Black RE, Allen LH, Bhutta ZA, Caulfield LE, de Onis M, Ezzati M, Mathers C, Rivera J (2008) Maternal and child undernutrition: global and regional exposures and health consequences. Lancet 371:243-260
19. Boualaphanh C, Calingacion M, Jothityangkoon D, Sanitchon J, Cuevas RM, Fitzgerald M (2011) Yield and quality of non-aromatic and aromatic Lao rice in response to nitrogen fertilizer. Science Asia 37:89-97
20. Boualaphanh C, Daygon D, Calingacion M, Jirawat , Jongithoon D, Mumm R, Hall R, Fitzgerald M (2011) Use of new generation SNP genotyping for rapid development of near-isogenic lines in

rice. Crop Science 51:

21. Bouis H, Hunt J (1999) Linking food and nutrition security: past lessons and future opportunities. Asian Developement Review 17:168-213

22. Bounphanousay C (2007) Use of phenotypic characters and DNA profiling for classification of the genetic diversity in black glutinous rice of the Lao PDR. Agriculture. Khon Kaen University. Ph. D. Agronomy, Khon Kaen. p 118

23. Brackmann C, Bengtsson A, Alminger ML, Svanberg U, Enejder A (2011) Visualization of -carotene and starch granules in plant cells using CARS and SHG microscopy. Journal of Raman Spectroscopy 42:586-592

24. Brand-Miller JC, Pang E, Bramall L (1992) Rice: a high or low glycemic index food? Am J Clin Nutr 56:1034-1036

25. Bryant RJ, McClung AM (2011) Volatile profiles of aromatic and non-aromatic rice cultivars using SPME/GC-MS. Food Chem 124:501-513

26. Butardo VM, Fitzgerald MA, Bird AR, Gidley MJ, Flanagan BM, Larroque O, Resurreccion AP, Laidlaw HK, Jobling SA, Morell MK, Rahman S (2011) Impact of Down-regulation of Starch Branching Enzyme IIb in Rice by Artificial microRNA and Hairpin-RNA-Mediated RNA Silencing. J Exp Bot 91:1298-1303

27. Calingacion M, Boualaphan C, Daygon D, Anacleto R, Biais B, Deborde C, Maucourt M, Moing A, Mumm R, Ric De Vos CH, Erban A, Hansen T, Laursen K, Shoerring J, Kopka J, Hall R, Fitzgerald M (2011) A metabolomics approach to identify new traits of rice quality in traditional and improved varieties of Laos. Metabolomics.

28. Champagne ET, Bett-Garber KL, Thomson JL, Shih FF, Lea J, Daigle K (2008) Impact of presoaking on flavor of cooked rice. Cereal Chemistry 85:706-710

29. Champagne ET, Bett-Garber KL, Fitzgerald MA, Grimm C, Lea J, Ohtsubo K, Jongdee S, Xie L, Bassinello P, Resurreccion AP, Ahmad R, Habibi F, Reinke R (2010) Important sensory properties differentiating premium rice varieties. Rice 3:270-281

30. Chen JT, Wesley R, Shamburek RD, Pucino F, Csako G (2005) Meta-analysis of natural therapies for hyperlipidemia: plant

sterols and stanols versus policosanol. Pharmacotherapy 25:171-183

31. Chen PN, Kuo WH, Chiang CL, Chiou HL, Hsieh YS, Chu SC (2006) Black rice anthocyanins inhibit cancer cells invasion via repressions of MMPs and u-PA expression. Chem Biol Interact 163:218-229
32. Chen H, He H, Zou Y, Chen W, Renbo Y, Liu X, Yang Y, Gao Y-M, Jian-Long X, Fan L-M, Li Y, Li Z-K, Deng X (2011) Development and application of a set of breeder-friendly SNP markers for genetic analyses and molecular breeding of rice (<i>Oryza sativa</i> L.). TAG Theoretical and Applied Genetics 123:869-879
33. Chukwu O (1999) Parboiling of rice paddy with heated pebbles. Journal of Science, Technology and Mathematics Education 2:70-76
34. Cicero AFG, Derosa G (2005) Rice bran and its main components: potencial role in the management of coronary risk factors. Current Topics in Nutraceutical Research 3:29-46
35. Curtis KR, McCluskey JJ, Wahl TI (2004) Consumer acceptance of genetically modified food products in the developing world. The Journal of Agrobiotechnology and Economics 7:70-75
36. Davy BM, Melby CL, Beske SD, Ho RC, Davrath LR, Davy KP (2002) Oat consumption does not affect resting casual and ambulatory 24-h arterial blood pressure in men with high-normal blood pressure to stage I hypertension. J Nutr 132:394-398
37. Eggum BO (1979) The nutritional value of rice in comparison with other cereals. Workshop on chemical aspects of rice grain quality, International Rice Research Institute, Los Banos, Laguna, Philippines.
38. FDA (1999) Health claim notification for whole grain foods.
39. Fernie AR, Schauer N (2009) Metabolomics-assisted breeding: a viable option for crop improvement? Trends Genet 25:39-48
40. Fernie , Alisdair R, Joost J, Keurentjes B (2011) Genetics, genomics and metabolomics. In Annual Plant Reviews 43:219-259 Wiley-Blackwell
41. Fidler IJ (2003) The pathogenesis of cancer metastasis: the 'seed and soil' hypothesis revisited. Nat Rev Cancer 3:453-458

42. Fiehn O, Kopka J, Dörmann P, Altmann T, Trethewey RN, Willmitzer L (2000) Metabolite profiling for plant functional genomics. Nat Biotechnol 18:1157-1161

43. Finney PL (1983) Effect of germination on cereal and legume nutrient changes and food and feed value: a comprehensive review. Plenum Press, New York and London, Recent advances in phytochemistry.

44. Fitzgerald M (2007) Screening the International Rice Genebank Collection for Variation in Carotenoid Content Harvestplus and IRRI. 5

45. Fitzgerald MA, McCouch SR, Hall RD (2009) More than just a grain of rice, the global quest for quality. Trends Plant Sci 14:133-139

46. Fitzgerald M, Concepcion J, Rahman S, Resurreccion A, Bird AR, Morell MK (2011) Identification of a major genetic determinant of glycaemic index in rice. Rice 4:66-74

47. Flight I, Clifton P (2006) Cereal grains and legumes in the prevention of coronary heart disease and stroke: A review of the literature. Eur J Clin Nutr 60:1145-1159

48. Foster-Powell K, Holt SHA, Brand-Miller JC (2002) International table of glycemic index and glycemic load values: 2002. Am J Clin Nutr 76:5-56

49. Fujino Y, Kuwata J (2004) Food functionality of sprout rice grain, super rice. Proceedings of the 12th International ICC Cereal and Bread Congress, Harrogate, UK., ICC.

50. Gariboldi F (1973) Rice testing methods and equipment. Food and Agriculture Organization of the United Nations, Rome.

51. Gerhardt AL, Gallo NB (1998) Full-fat rice bran and oat bran similarly reduce hypercholesterolemia in humans. J Nutr 128:865-869

52. Ghaffar A, Reddy KS, Singhi M (2004) Burden of non-communicable diseases in South Asia. Br Med J 328:807-810

53. Goffman FD, Bergman C (2002) Relationship between hydrolytic rancidity, oil concentration and esterase activity in rice bran. Cereal Chemistry 80:689-692

54. Goffman FD, Bergman CJ (2004) Rice kernel phenolic content and its relationship with antiradical efficiency. J Sci Food Agric 84:1235-1240

55. Goffman F, Pinson S, Bergman C (2003) Genetic diversity for lipid content and fatty acid profile in rice bran. J Am Oil Chem Soc 80:485-490

56. Gregorio GB (2002) Progress in breeding for trace minerals in staple crops. J Nutr 132:500S-502S

57. Grimm CG, Champagne ET, Lloyd SW, Easson M, Condon B, McClung AM (2011) Analysis of 2-Acetyl-1-pyrolline in rice by HSSE/GC/MS. Cereal Chemistry 88:271-277

58. Grusak MA (2005) Golden Rice gets a boost from maize. Nat Biotech 23:429-430

59. Haas JD, Beard JL, Murray-Kolb LE, del Mundo AM, Felix A, Gregorio GB (2005) Iron-Biofortified Rice Improves the Iron Stores of Nonanemic Filipino Women. J Nutr 135:2823-2830

60. Hall RD (2006) Plant metabolomics: from holistic hope, to hype, to hot topic. New Phytol 169:453-468

61. Hall RD (2011) Plant Metabolomics in a nutshell: potential and future challenges. In: Hall RD (ed) Biology of Plant Metabolomics, Wiley Blackwell, Oxford. pp 1-24

62. Hall RD, Brouwer ID, Fitzgerald MA (2008) Plant metabolomics and its potential application for human nutrition. Physiol Plant 132:162-175

63. Hallfrisch J, Daniel J, Scholfield , Kay M, Behall (2003) Blood pressure reduced by whole grain diet containing barley or whole wheat and brown rice in moderately hypercholesterolemic men. Nutr Res 23:1631-1642

64. Halton TL, Liu S, Manson JoAnn E, Frank BHu (2008) Low-carbohydrate-diet score and risk of type 2 diabetes in women. Am J Clin Nutr 87:339-346

65. Han SJ, Ryu S, Noh KSS (2004) A new 2-arylbenzofuran with antioxidant activity from the black colored rice (Oryza sativa L.) bran. Chem Pharm Bull 52:1365-1366

66. Hansen TH, Lombib E, Fitzgerald MA, Husted S, Boualaphanh C, Resurreccion A, Paterson D, Schjoerring JK (submitted) Losses of essential mineral nutrients by polishing of rice differ among genotypes due to contrasting grain hardness and mineral distribution. Cerea Chemistry.

67. Hawkes C (2004) Nutrition Labels and Health Claims: The Global Regulatory Environment. WHO, Geneva.

68. He L, Mo H, Hadisusilo S, Asaf A, Qureshi , Elson CE (1997) Isoprenoids suppress the growth of murine B16 melanomas In vitro and In vivo. J Nutr 127:668-674

69. Hegsted M, Windhauser MM, Kay Morris S, Lestera SB (1993) Stabilized rice bran and oat bran lower cholesterol in humans. Nutr Res 13:387-398

70. Heinemann RJB, Fagundes PL, Pinto EA, Penteado MVC, Lanfer-Marque UM (2005) Comparative study of nutrient composition of commercial brown, parboiled and milled rice from Brazil. Journal of Food Composition and Analysis 18:287-296

71. Heuberger AL, Lewis MR, Chen M-H, Brick MA, Leach JE, Ryan EP (2010) Metabolomic and functional genomic analyses reveal varietal differences in bioactive compounds of cooked rice. PLoS One 5:e12915

72. Hirunpong P, Tungjaroenchai W (2008) Effect of Germination on Contents of Bioactive Components in Germinated Brown Rice of Three Rice Cultivars. 34th Congress on Science and Technology, Thailand.

73. Hodge AM, English DR, O'Dea K, Giles GG (2004) Glycemic index and dietary fiber and the risk of type 2 diabetes. Diabetes Care 27:2701-2706

74. Hossain P, Kawar B, El NM (2007) Obesity and diabetes in the developing world – a growing challenge. N Engl J Med 356:213-215

75. Hotz C, Brown KH (2004) Assessment of the risk of zinc deficiency in populations and options for its control. Food Nutr Bull 25:S94-S203

76. Hudson E, Dinh P, Kokubun T, Simmonds M, Gescher A (2000) Characterization of potentially chemopreventive phenols in extracts of brown rice that inhibit the growth of human breast and colon cancer cells. Cancer Epidemiol Biomarkers Prev 9:1163-1170

77. Huffman WE (2004) Production, identity preservation, and labeling in a marketplace with genetically modified and non-genetically modified foods. Plant Physiol 134:3-10

78. Hunt JR, Johnson LK, Juliano BO (2002) Bioavailability of zinc from cooked Philippine milled, undermilled, and brown rice, as assessed in rats using growth, bone zinc, and zinc-65 retention. J Agric Food Chem 50:5229-5235

79. Jacobs DR, Andersen LF, Blomhoff R (2007) Whole-grain consumption is associated with a reduced risk of noncardiovascular, noncancer death attributed to inflammatory diseases in the Iowa women's health study. Am J Clin Nutr 85:1606-1614

80. James WPT (2008) The fundamental drivers of the obesity epidemic. Obes Rev 9:6-13

81. Jamison DT, Mosley HW, Measham AR, Bobadilla JL (1993) Disease control priorities in developing countries. World Bank, Washington (DC).

82. Jiang SL, Wu JG, Feng Y, Yang XE, Shi CH (2007) Correlation analysis of mineral element contents and quality traits in milled rice (Oryza sativa L.). J Agric Food Chem 55:9608-9613

83. Johnson AT, Kyriacou B, Callahan DL, Carruthers L, Stangoulis J, Lombi E, Tester M (2011) Constitutive overexpression of the OsNAS Gene family reveals single-gene strategies for effective iron- and zinc-biofortification of rice endosperm. PLoS One 6(9):e24476

84. Juliano BO (1985) Rice properties and processing. Food Reviews International 1:432-445

85. Kahlon TS, Chow FI, Sayre RN, Betschart AA (1991) Cholesterol-lowering in hamsters fed rice bran at various levels, defatted rice bran and rice bran oil. American Institute of Nutrition. pp 513-519

86. Kayahara H, Kikuichi T (2000) Flavor. Health and Nutritional Quality of Pregerminated Brown Rice, International Chemical Congress of Pacific Basin Societies, Hawaii.

87. Kind T, Scholz M, Fiehn O (2009) How large is the metabolome? A critical analysis of data exchange practices in chemistry. PLoS One 4:e5440

88. King RA, Noakes M, Bird AR, Morell MK, Topping DL (2008) An extruded breakfast cereal made from a high amylose barley cultivar has a low glycemic index and lower plasma insulin response than one made from a standard barley. Journal of Cereal Science 48:546-530

89. Kopelman PG (2000) Obesity as a medical problem. Nature 404:635-643

90. Kusano M, Tabuchi M, Fukushima A, Funayama K, Diaz C, Kobayashi M, Hatashi N, Tsuchiya YN, Takahashi H, Kamata A, Yamaya T, Saito K (2011) Metabolomics data reveal a crucial role of cystolic glutamine synthetase 1;1 in coordinating metabolic balance in rice. Plant J 66:456-466

91. Lamberts L, De Bie E, Derycke V, Veraverbeke WS, De Man W, Delcour JA (2006) Effect of processing conditions on color change of brown and milled parboiled rice. Cereal Chemistry 83:80-85

92. Leal J, Luengo-Fernández R, Gray A, Petersen S, Rayner M (2006) Economic burden of cardiovascular diseases in the enlarged European Union. Eur Heart J 27:1610-1619

93. Liang JFB, Han ZL, Han MJ, Nout R, Hamer RJ (2007) Iron, zinc and phytic acid content of selected rice varieties from China. J Sci Food Agric 87:504-510

94. Liang J, Li Z, Tsuji K, Kazuhiko Nakano MJ, Nout R, Hamer RJ (2008) Milling characteristics and distribution of phytic acid and zinc in long-, medium- and short-grain rice. Journal of Cereal Science 48:83-91

95. Liotta LA, Stetler-Stevenson WG (1991) Tumor invasion and metastasis: an imbalance of positive and negative regulation. Cancer Research (Supplement) 51:5054s-5059s

96. Liu JC, Ockenden I, Truax M, Lott JNA (2004) Phytic acid-phosphorus and other nutritionally important mineral nutrient elements in grains of wild-type and low phytic acid (lpa1-1) rice. Seed Science Research 14:109-116

97. Lochner J, Rugge B, Judkins D (2006) How effective are lifestyle changes for controlling hypertension? J Fam Pract 55:73-74

98. London S (2004) Cardiovascular disease threatens developing countries. Br Med J 328:1034

99. Lopez-Ridaura R, Willett WC, Rimm EB, Liu S, Stampfer MJ, Manson JE, Hu FB (2004) Magnesium intake and risk of type 2 diabetes in men and women. Diabetes Care 27:134-140

100. Lucca P, Hurrell R, Potrykus I (2002) Fighting iron deficiency anemia with iron-rich rice. J Am Coll Nutr 21:184S-190S

101. Ma G, Jin Y, Piao J, Kok F, Guusje B, Jacobsen E (2005) Phytate, calcium, iron, and zinc contents and their molar ratios in foods commonly consumed in China. J Agric Food Chem 53:10285-10290

102. Manful JT, Swetman AA, Coker RD, Drunis A (2007) Changes in the thiamine and riboflavin contents of rice during artisanal parboiling in Ghana. Tropical Science 47:211-217

103. Mason J, Bailes A, Beda-Andourou M, Copeland N, Curtis T, Deitchler M, Foster L, Hensley M, Horjus P, Johnson C, Lloren T, Mendez A, Munoz M, Rivers J, Vance G (2005) Recent trends in malnutrition in developing regions: Vitamin A deficiency, anemia, iodine deficiency, and child underweight. Food Nutr Bull 26:59-107

104. McCouch SR, Teytelman L, Xu YB, Lobos KB, Clare K, Walton M, Fu BY, Maghirang R, Li ZK, Xing YZ, Zhang QF, Kono I, Yano M, Fjellstrom RG, DeClerck G, Schneider D, Cartinhour S, Ware D, Stein L (2002) Development and mapping of 2240 new SSR markers for rice (Oryza sativa L.). DNA Res 9:199-207

105. McCouch SR, Zhao K, Wright M, Tung C-W, Ebana K, Thomson MJ, Reynolds A, Wang D, Geneviev deClerck M, Liakat A, Anna M, McClung Georgia E, Carlos D, Bustamante (2010) Development of genome-wide SNP assays for rice. Breeding Science 60:524-535

106. Misra A, Singhal N, Khurana L (2010) Obesity, the metabolic syndrome, and type 2 diabetes in developing countries: Role of dietary fats and oils. J Am Coll Nutr 29:289S-301S

107. Most MM, Tulley R, Morales S, Lefevre M (2005) Rice bran oil, not fiber, lowers cholesterol in humans. Am J Clin Nutr 81:64-68

108. Munter de Jeroen SL, Frank BHu, Franz DSM, van Rob M, Dam (2007) Whole grain, bran, and germ Intake and risk of type 2 diabetes: a prospective cohort study and systematic review. PLoS Med 4:261

109. Murray-Kolb LE, Takaiwa F, Goto F, Yoshihara T, Theil EC, Beard JL (2002) Transgenic rice is a source of iron for iron-depleted rats. J Nutr 132:957-960

110. Nakamura H (1966) Effect of gamma-oryzanol on hepatic cholesterol biosynthesis and faecal excretion of cholesterol metabolites. Radioisotopes 15:371-374

111. Nanri A, Mizoue T, Noda M, Takahashi Y, Kato M, Inoue M, Tsugane S (2010) Rice intake and type 2 diabetes in Japanese men and women: the Japan Public Health Center-based Prospective Study. Am J Clin Nutr 92:1468-1477

112. Nesaretnam K, Meganathan (2011) Tocotrienols: inflammation and cancer. Annals of the New York Acadamy of Science 1229:18-22

113. Nicolosi RJ, Austrian LM, Mark Hegsted D (1991) Rice bran oil lowers serum total and low density lipoprotein cholesterol and apo B levels in nonhuman primates. Atherosclerosis 88:133-142

114. Nielsen MM, Hansen Å (2008) Rapid high performance liquid chromatography determination of tocopherols and tocotrienols in cereals. Cereal Chemistry 85:248-251

115. Nisak MY, Barakatun AR, Talib AK, Norimah HG, Azmi KN (2010) Improvement of dietary quality with the aid of a low glycemic index diet in Asian patients with type 2 diabetes mellitus. J Am Coll Nutr 29:161-170

116. Nugent R (2008) Chronic diseases in developing countries. Ann N Y Acad Sci 1136:70-79

117. Nunes SG, Gomes JC, Cruz R, Jordan CP (1991) Mineral enrichment of rice with hydrothermal treatments/mineral enrichement hidrothermal during processing of rice. Brazilian Archives of Biology and Technology 34:571-582

118. Oikawa A, Matsuda F, Kusano M, Okazaki Y, Saito K (2008) Rice metabolomics. Rice 1:63-71

119. Oki T, Masuda M, Nagai S, Take'ichi M, Kobayashi M, Nishiba Y, Sugawara T, Suda I, Sato T (2004) Radical-scavenging activity of red and black rice. Rice is life: scientific perspectives for the 21st century. Proceedings of the World Rice Research Conference Los Banos (Philippines). International Rice Research Institute and Tsukuba (Japan): Japan International Research Center for Agricultural Sciences, Los Banos, Philippines.

120. Otegbayo BO, Osamuel F, Fashakin JB (2001) Effect of parboiling on physico-chemical qualities of two local rice varieties in nigeria. The Journal of Food Technology in Africa 6:130-133

121. Packer L, Weber SU, Rimbach G (2001) Molecular aspects of alpha-tocotrienol antioxidant action and cell signalling. J Nutr 131:369S-373S

122. Panchan K, Naivikul O (2009) Effect of pre-germination and parboiling on brown rice properties. Asian Journal of Food Agriculture Industries 2:515-524

123. Panlasigui LN, Thompson LU (2006) Blood glucose lowering effects of brown rice in normal and diabetic subjects. International Journal of Food Science and Nutrition 57:151-158

124. Parashar UD, Hummelman EG, Bresee JS, Miller MA, Glass RI (2003) Global illness and deaths caused by rotavirus disease in children. Emerging Infectious Diseases journal 9:565-572

125. Park Y, David J, Hunter DS, Bergkvist L, Berrino F, van den Piet A, Brandt JE, Buring GA, Colditz Jo L, Freudenheim CS, Fuchs EG, Alexandra Goldbohm R, Graham S, Harnack L, Anne M, Hartman DR, Jacobs IK, Krogh V, Michael F, Leitzmann ML, McCullough AB, Miller PP, Thomas E, Rohan AS, Walter C, Willett AW, Zeleniuch-Jacquotte A, Shumin M, Zhang , Stephanie A, Smith-Warner (2005) Dietary fiber intake and risk of colorectal cancer. J Am Med Assoc 294:2849-2857

126. Parker RA, Pearce BC, Clark RW, Gordon DA, Wright JJ (1993) Tocotrienols regulate cholesterol production in mammalian cells by post-transcriptional suppression of 3-hydroxy-3-methylglutaryl-coenzyme A reductase. J Biol Chem 268:11230-11238

127. Pedersen B, Knudsen KE, Eggum BO (1989) Nutritive value of cereal products with emphasis on the effect of milling. World Rev Nutr Diet 60:1-91

128. Pena B, Yelitza J, Frank P, Mario T (2008) Zinc sérico en menores de 15 años de una comunidad rural del estado Lara. Annal Venezuelan Nutrition 21:77-84

129. Pins JJ, Geleva D, Keenan JM, Frazel CO, Connor PJ, Cherney LM (2002) Do whole-grain oat cereals reduce the need for antihypertensive medications and improve blood pressure control? J Fam Pract 51:353-359

130. Prasad AS (2003) Zinc deficiency. Br Med J 326:409-410

131. Prasad AS, Miale AJ, Farid Z, Sandstead H, Schulert A (1963) Zinc metabolism in patients with syndrome of iron deficiency anemia, hepatosplenomyaly, dwarfism and hypogonadism. J Lab Clin Med 61:537-549

132. Qu LQ, Yoshihara T, Ooyama A, Goto F, Takaiwa F (2005) Iron accumulation does not parallel the high expression level of ferritin in transgenic rice seeds. Planta 222:225-233

133. Qureshi AA, Basil A, Bradlow WA, Salser L, Brace D (1997) Novel tocotrienols of rice bran modulate cardiovascular disease risk parameters of hypercholesterolemic humans. J Nutr Biochem 8:290-298

134. Qureshi AA, Sami SA, Salser WA, Khan FA (2002) Dose-dependent suppression of serum cholesterol by tocotrienol-rich fraction (TRF25) of rice bran in hypercholesterolemic humans. Atherosclerosis 161:199-207

135. Ramakrishnan TV, Francis FJ (1979) Stability of carotenoids in model aqueous systems. Journal of Food Quality 2:177-189

136. Ranawana DV, Henry CJK, Lightowler HJ, Wang D (2009) Glycaemic index of some commercially available rice and rice products in Great Britain. Int J Food Sci Nutr 60:99-110

137. Rasool AHG, Wong AR (2007) Tocotrienol rich vitamin E: A review of clinical studies. International Medical Journal 14:139-143

138. Reiner Z, Tedeschi-Reiner E, Romi Z (2005) Effects of rice policosanol on serum lipoproteins, homocysteine, fibrinogen and C-reactive protein in hypercholesterolaemic patients. Clin Drug Investig 25:701-707

139. Rukmini C, Raghuram TC (1991) Nutritional and biochemical aspects of the hypolipidemic action of rice bran oil: a review. J Am Coll Nutr 10:593-601

140. Ryan EP, Heuberger AL, Weir TL, Barnett B, Broeckling CD, Prenni JE (2011) Rice bran fermented with Saccharomyces boulardii generates novel metabolite profiles with bioactivity. J Agric Food Chem 59:1862-1870

141. Sakamoto S, Hayashi T, Hayashi K, Murai F, Hori M, Kimoto K, Murakami K (2007) Pre-germinated brown rice could enhance

maternal mental health and immunity during lactation. Eur J Nutr 46:391-396

142. Sato S, Soga T, Nishioka T, Tomita M (2004) Simultaneous determination of the main metabolites in rice leaves sing capillary electrophoresis mass pectrometry and capillary electrophoresis diode array etection. Plant J 40:151-163
143. Sato S, Arita M, Soga T, Nishioka T, Tomita M (2008) Time-resolved metabolomics reveals metabolic modulation in rice foliage. BMC Syst Biol 2:51-64
144. Sato S, Fukumura K, Nishiyama A, Kanamoto I, Inoue Y, Konishi T (2010) Glycemic index and glucose utilization of rice vermicelli in healthy subjects. Biol Pharm Bull 33:1385-1393
145. Schaffer S, Pallauf J, Krawinkel MB (2004) Impact of feeding high-iron rice on plasma iron, hemoglobin and red blood cell variables of early-weaned piglets—A pilot study. Ann Nutr Metab 48:109-117
146. Schatzkin A, Mouw T, Park Y, Amy F, Subar VK, Hollenbeck A, Michael F, Leitzmann F, Thompson E (2007) Dietary fiber and whole-grain consumption in relation to colorectal cancer in the NIH-AARP diet and health study. Am J Clin Nutr 85:1353-1360
147. Seal CJ, Jones AR, Whitney AD (2006) Whole grains uncovered. Nutrition Bulletin 31:129-137
148. Sen CK, Khanna S, Roy S (2007) Tocotrienols in health and disease: the other half of the natural vitamin E family. Mol Aspects Med 28:692-728
149. Shallan MA, El-Beltagi HS, Mona AM, Amera TM (2010) Chemical evaluation of pre-germinated brown rice and whole grain rice bread. Electronic Journal of Environmental, Agricultural and Food Chemistry 9(3):958-971
150. Shaw JE, Sicree R, Zimmet P (2010) Global estimates of the prevalence of diabetes for 2010 and 2030. Diabetes Res Clin Pract 87:4-14
151. Shaw , Julia G, Jennifer F, Friedman (2011) Iron deficiency anemia: focus on infectious diseases in lesser developed countries. Anemia, Review Article 2011:10

152. Shen Y, Jin L, Xiao P, Yan L, Bao J-S (2009) Total phenolics, flavonoids, antioxidant capacity in rice grain and their relations to grain color, size and weight. Journal of Cereal Science 49:106-111

153. Lee S, Jeon Un S, Lee Seung J, Kim Y-K, Persson Daniel P, Soren H, Schjørring Jan K, Kakei Y, Masuda H, Nishizawa Naoko K, An G (2009) Iron fortification of rice seeds through activation of the nicotianamine synthase gene. Proceedings of the National Academy of Sciences 106:22014-22019

154. Shimizu T (2011) Health claims and scientific substantiation of functional foods – Japanese regulatory system and the international comparison. European Food and Feed Law Review 6:144-152

155. Slavin JL, Martini MC, Jacobs DR, Marquart L (1999) Plausible mechanisms for the protectiveness of whole grains. Am J Clin Nutr 70:459S-463S

156. Statistics (2006) Non-Communicable Diseases Surveillance. Malaysia.

157. Statistics (2010) Statistics On Causes Of Death. Malaysia.

158. Stewart D, Louise V, Shepherd T, Robert D, Hall P, Fraser D (2011) Crops and tasty, nutritious food – how can metabolomics help? In Annual Plant Reviews 43:181-217 Wiley-Blackwell

159. Storck CR, da Silva LP, Fagundes CAA (2005) Categorizing rice cultivars based on differences in chemical composition. Journal of Food Composition and Analysis 18:333-341

160. Sun Q, Spiegelman D, van Dam RM, Holmes MD, Malik VS, Willett WC, Hu FB (2010) White rice, brown rice, and risk of type 2 diabetes in US men and women. Arch Intern Med 170:961-969

161. Sundstrom FJ, Williams J, Van A, Deynze , Bradford KJ (2002) Identity preservation of agricultural commodities. Agricultural Biotechnology in California Series 8077:15

162. Sung HG, Shin HT, Ha JK, Lai HL, Cheng KJ, Lee JH (2005) Effect of germination temperature on characteristics of phytase production from barley. Bioresour Technol 96:1297-1303

163. Tamura T, Goldenberg RL (1996) Zinc nurture and pregnancy outcome. Nutr Res 16:139-181

164. Tang G, Qin J, Dolnikowski GG, Russel RM, Grusak MA (2009) Golden Rice is an effective source of vitamin A. Am J Clin Nutr 89:1776-1783

165. Tarpley L, Roessner U (2007) Metabolomics: enabling systems-level phenotyping in rice functional genomics. In: Upadhyaya NM (ed) Rice Functional Genomics: Challenges, Progress and Prospects, Springer, New York. pp 91-107

166. Tian S, Nakamura K, Kayahara H (2004) Analysis of phenolic compounds in white rice. Brown Rice, and Germinated Brown Rice Journal of Agricultural & Food Chemistry 15:4808-4813

167. Topping DL (2007) Cereal complex carbohydrates and their contribution to human health. Journal of Cereal Science 46:220-229

168. Topping DL, Clifton PM (2001) Short-chain fatty acids and human colonic function: Roles of resistant starch and nonstarch polysaccharides. Physiol Rev 81:1031-1064

169. Topping D, Fukushima M, Bird AR (2003) Resistant starch as a prebiotic and synbiotic: state of the art. Proc Nutr Soc 62:171-176

170. Tucker G (2003) Nutritional enhancement of plants. Curr Opin Biotechnol 14:221-225

171. Tung C-W, Zhao K, Mark Wright M, Ali JJ, Kimball J, Tyagi W, Thomson M, McNally K, Leung H, Kim H, Ahn S-N, Reynolds A, Scheffler B, Eizenga G, McClung A, Bustamante C, McCouch S (2011) Development of a research platform for dissecting phenotypeâ€"genotype associations in rice (<i > Oryza</i > spp.). Rice 3:205-217

172. Udomkesmalee E, Dhanamitta S, Yhoung-Aree J, Rojroongwasinkul N, Smith JC (1990) Biochemical evidence suggestive of suboptimal zinc and vitamin A status in schoolchildren in Northeast Thailand. American Journul of Clinical Nutrition 52:564-567

173. Usuki SY, Ito Y, Morikawa K, Kise M, Toyohiko A, Rivner M, Yu R (2007) Effect of pre-germinated brown rice intake on diabetic neuropathy in streptozotocin-induced diabetic rats. Nutr Metab 4:1-11

174. Varady KA, Wang Y, Jones PJ (2003) Role of policosanols in the prevention and treatment of cardiovascular disease. Nutr Rev 61:376-383

175. Vasconcelos M, Datta K, Oliva N, Khalekuzzaman M, Torrizo L, Krishnan S, Oliveira M, Goto F, Datta SK (2003) Enhanced iron and zinc accumulation in transgenic rice with the ferritin gene. Plant Sci 164:371-378

176. Verhoeven HA, Jonker H, DeVos RCH, Hall RD (2011) Measuring plant volatiles. In: Hardy NG, Hall RD (eds) Methods for Plant metabolomics, Humana Press.

177. Verschoyle RD, Greaves P, Cai H, Edwards RE, Steward WP, Gescher AJ (2007) Evaluation of the cancer chemopreventive efficacy of rice bran in genetic mouse models of breast, prostate and intestinal carcinogenesis. Br J Cancer 96:248-254

178. Villegas R, Liu S, Gao YT, Yang G, Li H, Zheng W, Shu XO (2007) Prospective study of dietary carbohydrates, glycemic index, glycemic load, and incidence of type 2 diabetes mellitus in middle-aged Chinese women. Arch Intern Med 26:2310-2316

179. Virk P, Gerard B (2007) Biofortified rice – towords combating human micronutrient deficiencies. International Rice Research Institute, DAPO Box 7777, Metro Manila, Philippines.

180. Wang RM, Yang XE, He CX (2004) Genetic analysis on agronomic traits related to zinc efficiency in lowland rice. Journal of Plant Nutrition 27:585-599

181. Wang L, Michael Gaziano J, Liu S, JoAnn E, Manson JE, Howard B, Sesso D (2007) Whole- and refined-grain intakes and the risk of hypertension in women. Am J Clin Nutr 86:472-479

182. Welch RM, Graham RD (2004) Breeding for micronutrients in staple food crops from a human nutrition perspective. J Exp Bot 55:353-364

183. West KP (2002) Extent of vitamin A deficiency among preschool children and women of reproductive age. J Nutr 132:2857S-2866S

184. WHO (2001) World health report 2001, health systems: improving performance. World Health Organization, Geneva.

185. WHO (2002) Health situation in the South East Asia Region 1998–2000. Regional Office for South East Asia, World Health Organization, New Delhi. p p 2002

186. WHO (2008) Worldwide prevalence of anaemia 1993–2005. World Health Organization, 20 Avenue Appia, 1211, Geneva 27, Switzerland.

187. Wild S, Roglic G, Green A, Sicree R, King H (2004) Global prevalence of diabetes. Diabetes Care 27:1047-1053

188. Williams MT, Hord NG (2005) The role of dietary factors in cancer prevention: beyond fruits and vegetables. Nutr Clin Pract 20:451-459

189. Williams P, Brand Miller J, Fitzgerald M (2005) Project 4505 GI of Rice. Cooperative Research Centre for Sustainable Rice Production, Leeton, Australia. pp 84-89

190. Wilson TA, Nicolosi RJ, Woolfrey B, Kritchevsky D (2007) Rice bran oil and oryzanol reduce plasma lipid and lipoprotein cholesterol concentrations and aortic cholesterol ester accumulation to a greater extent than ferulic acid in hypercholesterolemic hamsters. J Nutr Biochem 18:105-112

191. Wright M, Tung C-W, Zhao K, Reynolds A, McCouch SR, Bustamante CD (2010) ALCHEMY: a reliable method for automated SNP genotype calling for small batch sizes and highly homozygous populations. Bioinformatics (Oxford) 26:2952-2960

192. Xu Z, Samuel Godber J (1999) Purification and identification of components of gamma-oryzanol in rice bran oil. J Agric Food Chem 47:2724-2728

193. Yamakawa H, Hakata M (2010) Atlas of rice grain filling-related metabolism under high temperature: Joint analysis of metabolome and transcriptome demonstrated inhibition of starch accumulation and induction of amino acid accumulation. Plant Cell Physiol 51:795-809

194. Yoon KH, Lee JH, Kim JW, Cho JH, Choi YH, Ko SH, Zimmet P, Son HY (2006) Epidemic obesity and type 2 diabetes in Asia. Lancet 368:1681-1688

195. You D, Wardlaw T, Salama P, Jones G (2010) Levels and trends in under-5 mortality, 1990–2008. Lancet 375:100-103

196. Yunus AM, Sherina MS, Nor Afiah MZ, Rampal L, Tiew KH (2004) Prevalence of cardiovascular risk factors in a rural Community in Mukim Dengkil, Selangor. Malaysian Journal of Nutrition 10:5-11

197. Zaidi AKM, Awasthi S, Janaka H, De S (2004) Burden of infectious diseases in South Asia. Br Med J 328:811-815

198. Zhang G, Vasanti Malik S, Pan A, Kumar S, Holmes M, Donna Spiegelman X, Lin , Frank Hu B (2010) Substituting brown rice for white rice to lower diabetes risk: A focus-group study in Chinese adults. J Am Diet Assoc 110:1216-1221
199. Zheng L, Cheng Z, Ai C, Jiang X, Bei X, Zheng Y, Glahn RP, Welch RM, Miller DD, Lei XG, Shou H (2010) Nicotianamine, a novel enhancer of rice iron bioavailability to humans. PLoS One 5:e10190
200. Zhou JM, Ibrahim RK (2010) Tricin—a potential multifunctional nutraceutical. Phytochemistry Reviews 9:413-424
201. Zhou Z, Robards K, Helliwell S, Blanchard C (2004) The distribution of phenolic acids in rice. Food Chem 87:401-406
202. Zhou J, Ma C, Xu H, Yuan KL, Lu X, Zhu Z, Wu YN, Xu GW (2008) Metabolic profiling of transgenic rice with cryIAc and sck genes: an evaluation of unintended effects at metabolic level by using GC-FID and GC-MS. J Chromatogr B 877:725-732

Chapter 7

Traditional Medicines in Africa: An Appraisal of Ten Potent African Medicinal Plants

M. Fawzi Mahomoodally

Department of Health Sciences, Faculty of Science, University of Mauritius, 230 Réduit, Mauritius

ABSTRACT

The use of medicinal plants as a fundamental component of the African traditional healthcare system is perhaps the oldest and the most assorted of all therapeutic systems. In many parts of rural Africa, traditional healers prescribing medicinal plants are the most easily accessible and affordable health resource available to the local community and at times the only therapy that subsists. Nonetheless, there is still a paucity of updated comprehensive compilation of promising medicinal plants from the African continent. The major focus of the present review is to provide an updated overview of 10 promising medicinal plants from the African biodiversity which have short- as well as long-term potential to be developed as future phytopharmaceuticals to treat and/or manage panoply of infectious and chronic conditions. In this endeavour, key scientific databases have been probed to investigate trends in the rapidly increasing number of scientific publications on African traditional medicinal plants. Within the framework of enhancing the significance of traditional African medicinal plants, aspects such as traditional use, phytochemical profile, in vitro, in vivo, and clinical studies and also future challenges pertaining to the use of these plants have been explored.

INTRODUCTION

Traditional medicine is the sum total of knowledge, skills, and practices based on the theories, beliefs, and experiences indigenous to different cultures that are used to maintain health, as well as to prevent, diagnose, improve, or treat physical and mental illnesses [1]. Traditional medicine that has been adopted by other populations (outside its indigenous culture) is often termed complementary or alternative medicine (CAM) [1, 2].

The World Health Organization (WHO) reported that 80% of the emerging world's population relies on traditional medicine for therapy. During the past decades, the developed world has also witnessed an ascending trend in the utilization of CAM, particularly herbal remedies [3]. Herbal medicines include herbs, herbal materials, herbal preparations, and finished herbal products that contain parts of plants or other plant materials as active ingredients. While 90% of the population in Ethiopia use herbal remedies for their primary healthcare, surveys carried out in developed countries like Germany and Canada tend to show that at least 70% of their population have tried CAM at least once [2, 3]. It is likely that the profound knowledge of herbal remedies in traditional cultures, developed through trial and error over many centuries, along with the most important cures was carefully passed on verbally from one generation to another. Indeed, modern allopathic medicine has its roots in this ancient medicine, and it is likely that many important new remedies will be developed and commercialized in the future from the African biodiversity, as it has been till now, by following the leads provided by traditional knowledge and experiences [2–5].

The extensive use of traditional medicine in Africa, composed mainly of medicinal plants, has been argued to be linked to cultural and economic reasons. This is why the WHO encourages African member states to promote and integrate traditional medical practices in their health system [1]. Plants typically contain mixtures of different phytochemicals, also known as secondary metabolites that may act individually, additively, or in synergy to improve health. Indeed, medicinal plants, unlike pharmacological drugs, commonly have several chemicals working together catalytically and synergistically to produce a combined effect that surpasses the total activity of the individual constituents. The combined actions of these substances tend to increase the activity of the main medicinal constituent by speeding up or slowing down its assimilation in the body. Secondary metabolites from plant's origins might increase the stability of the active compound(s) or phytochemicals, minimize the rate of undesired adverse side effects, and have an additive, potentiating, or antagonistic effect. It has been postulated that the enormous diversity of chemical structures found in these plants is not waste products, but specialized secondary metabolites involved in the relationship of the organism with the environment, for example, attractants of pollinators, signal products, defensive substances against predators and parasites, or in resistance against pests and diseases. A single plant may, for example, contain bitter substances that stimulate digestion and possess anti-inflammatory compounds that reduce swellings and pain, phenolic compounds that can act as an antioxidant and venotonics, antibacterial and antifungal tannins that act as natural antibiotics, diuretic substances that enhance the elimination of waste products and toxins, and alkaloids that enhance mood and give a sense of well-being [1–5]. Although some may view the isolation of phytochemicals and their use as single chemical entities as a better alternative and which have resulted in the replacement of plant extracts' use, nowadays, a view that there may be some advantages of the medical use of crude and/or standardized extracts as opposed to isolated single compound is gaining much momentum in the scientific community.

AFRICAN TRADITIONAL MEDICINE

African traditional medicine is the oldest, and perhaps the most assorted, of all therapeutic systems. Africa is considered to be the cradle of mankind with a rich biological and cultural diversity marked by regional differences in healing practices [2, 6]. African traditional medicine in its varied forms is holistic involving both the body and the mind. The traditional healer typically diagnoses and treats the psychological basis of an illness before prescribing medicines, particularly medicinal plants to treat the symptoms [2, 6–8]. The sustained interest in traditional medicine in the African healthcare system can be justified by two major reasons. The first one is inadequate access to allopathic medicines and western forms of treatments, whereby the majority of people in Africa cannot afford access to modern medical care either because it is too costly or because there are no medical service providers. Second, there is a lack of effective modern medical treatment for some ailments such as malaria and/or HIV/AIDS, which, although global in distribution, disproportionately affect Africa more than other areas in the world.

The most common traditional medicine in common practice across the African continent is the use of medicinal plants. In many parts of Africa, medicinal plants are the most easily accessible health resource available to the community. In addition, they are most often the preferred option for the patients. For most of these people, traditional healers offer information, counseling, and treatment to patients and their families in a personal manner as well as having an understanding of their patient's environment [2, 6, 7]. Indeed, Africa is blessed with enormous biodiversity resources and it is estimated to contain between 40 and 45,000 species of plant with a potential for development and out of which 5,000 species are used medicinally. This is not surprising since Africa is located within the tropical and subtropical climate and it is a known fact that plants accumulate important secondary metabolites through evolution as a natural means of surviving in a hostile environment [9]. Because of her tropical conditions, Africa has an unfair share of strong ultraviolet rays of the tropical sunlight and numerous pathogenic microbes, including several species of bacteria, fungi, and viruses, suggesting that African plants could accumulate chemopreventive substances more than plants from the northern hemisphere. Interestingly, Abegaz et al. [10] have observed that of all species of Dorstenia(Moraceae) analysed, only the African species, Dorstenia mannii Hook.f, a perennial herb growing in the tropical rain forest of Central Africa contained more biological activity than related species [9–11].

Nonetheless, the documentation of medicinal uses of African plants and traditional systems is becoming a pressing need because of the rapid loss of the natural habitats of some of these plants due to anthropogenic activities and also due to an erosion of valuable traditional knowledge. It has been reported that Africa has some 216 million hectares of forest, but the African continent is also notorious to have one of the highest rates of deforestation in the world, with a calculated loss through deforestation of 1% per annum [7, 12]. Interestingly, the continent also has the highest rate of endemism, with the Republic of Madagascar topping the list by 82%, and it is worth to emphasize that Africa already contributes nearly 25% of the world trade in biodiversity. Nonetheless, the paradox is that in spite of this huge potential and diversity, the African continent has only few drugs commercialized globally [2, 12, 13].

The scientific literature has witnessed a growing number of publications geared towards evaluating the efficacy of medicinal plants from Africa which are believed to have an important contribution in the maintenance of health and in the introduction of new treatments. Nonetheless, there is still a dearth of updated comprehensive compilation of promising medicinal plants from the African continent.

The main aim of the present review is to highlight the importance and potential of medicinal plants from the African biodiversity which have short- as well as long-term potential to be developed as future phytopharmaceuticals to treat and/or manage panoply of infectious and chronic conditions. The review might also provide a starting point for future studies aimed at isolation, purification, and characterization of bioactive compounds present in these plants as well as exploring the potential niche market of these plants. In this endeavor, major scientific databases such as EBSCOhost, PubMed Central, Scopus (Elsevier), and Emerald amongst others have been probed to investigate trends in the rapidly increasing number of scientific publications on African traditional medicinal plants. Ten medicinal plants (Acacia senegal, Aloe ferox, Artemisia herba-alba, Aspalathus linearis, Centella asiatica, Catharanthus roseus, Cyclopia genistoides, Harpagophytum procumbens, Momordica charantia, andPelargonium sidoides) of special interest were chosen for more detailed reviews based on the following criteria: medicinal plants that form part of African herbal pharmacopeia with commercial importance and those plants from which modern phytopharmaceuticals have been derived.

Acacia Senegal **(L.) Willd. (Leguminosae: Mimosoideae)—Gum Arabic**

Acacia senegal, also known as gum Arabic, is native to semidesert and drier regions of sub-Saharan Africa, but widespread from Southern to Northern Africa. It is used as a medicinal plant in parts of Northern Nigeria, West Africa, North Africa, and other parts of the world [8]. The use of gum arabic (or gum acacia), which is derived from an exudate from the bark, dates from the first Egyptian Dynasty (3400 B.C.). It was used in the production of ink, which was made from a mixture of carbon, gum, and water. Inscriptions from the 18th Dynasty refer to this gum as "komi" or "komme." Gum arabic has been used for at least 4,000 years by local people for the preparation of food, in human and veterinary medicine, in crafts, and as a cosmetic. The gum of A. senegal has been used medicinally for centuries, and various parts of the plant are used to treat infections such as bleeding, bronchitis, diarrhea, gonorrhea, leprosy, typhoid fever, and upper respiratory tract infections. African herbalists use gum acacia to bind pills and to stabilize emulsions. It is also used in aromatherapy for applying essential oils [8, 14–16].

Currently, A. senegal is an important naturally occurring oil-in-water emulsifier, which is in regular use in the food and pharmaceutical industries. Medicinally, gum arabic is used extensively in pharmaceutical preparations and is a food additive approved as toxicologically safe by the Codex Alimentarius. It has been used as demulcent, skin protective agent, and pharmaceutical aids such as emulsifier and stabilizer of suspensions and additives for solid formulations. It is sometimes used to treat bacterial and fungal infections of the skin and mouth. It has been reported to soothe the mucous membranes of the intestines and to treat inflamed skin [17, 18]. The demulcent, emollient gum is used internally against inflammation of intestinal mucosa and externally to cover inflamed surfaces, as burns, sore nipples, and nodular leprosy. Additionally, it has also been documented to be used as antitussive, expectorant, astringent, catarrh and against colds, coughs, diarrhea, dysentery, gonorrhea, hemorrhage, sore throat, typhoid, and for urinary tract ailments [18]. The gum of A. senegal has been pharmaceutically used mainly in the manufacture of emulsions and in making pills and troches (as an excipient), as demulcent for inflammations of the throat or stomach and as masking agent for acrid tasting substances such as capsicum and also as a film-forming agent in peel-off masks. Gum arabic is also used widely as an ingredient in foods like candies and soft drinks as the gum has the properties of glue that is safe to eat. Gum acacia is widely used in organic products as natural alternative to chemical binders and is used in commercial emulsification for the production of beverages and flavor concentrates [8, 17–19].

Recently, it has been reported that A. senegal bark extracts were evaluated in vitro for their antimicrobial potential against human pathogenic isolates (Escherichia coli, Staphylococcus aureus, Streptococcus pneumoniae, Klebsiella pneumoniae, Shigella dysenteriae, Salmonella typhi, Streptococcus pyogenes,Pseudomonas aeruginosa, and Proteus vulgaris). The extract was found to exhibit significant antibacterial activity which was suggested to be due to the presence of tannins and saponins in the plant. It was also reported that the plant extract may not be toxic to man following in vitro cytotoxicity evaluation [18].

Aloe ferox **Mill. (Xanthorrhoeaceae)—Bitter Aloe or Cape Aloe**

Aloe ferox is native to South Africa and Lesotho and is considered to be the most common Aloe species in South Africa. A. ferox has been used since time immemorial and has a well-documented history of use as an alternative medicine and is one of the few plants depicted in San rock paintings. The bitter latex, known as Cape aloe, is used as laxative medicine in Africa and Europe and is considered to have bitter tonic, antioxidant, anti-inflammatory, antimicrobial, and anticancer properties [8, 19–23].

The use of A. ferox as a multipurpose traditional medicine has been translated into several commercial applications and it is a highly valued plant in the pharmaceutical, natural health, food, and cosmetic industries. A. ferox is considered South Africa's main wild harvested commercially traded species. The finished product obtained from aloe tapping, aloe bitters, has remained a key South African export product since 1761 when it was first exported to Europe. The aloe tapping industry is the livelihood of many rural communities and formalization of the industry in the form of establishment of cooperatives and trade agreements. It has been suggested that its trade may have an extensive poverty alleviation effect in Africa [19, 24, 25].

A. ferox has many traditional and documented medicinal uses. It is most popularly used for its laxative effect and as a topical application to the skin, eyes, and mucous membranes. Scientific studies conducted have verified many of the traditional uses. More recently, the cosmetic industry has shown interest in A. ferox gel [8, 19]. It has been reported that A. ferox gel contains at least 130 medicinal agents with anti-inflammatory, analgesic, calming, antiseptic, germicidal, antiviral, antiparasitic, antitumour, and anticancer effects encompassing all of the traditional uses and scientific studies done on A. feroxand its constituents [8, 24–31].

A wide variety of phenolic compounds (chromones, anthraquinones, anthrone, anthrone-C-glycosides, and other phenolic compounds) are present within A. ferox and these compounds have been well documented to possess biological activity [21, 24–31]. The gel polysaccharides are known to be of the arabinogalactan and rhamnogalacturonan types. The leaf gel composition still remains unknown and its claimed biological activities still remain to be investigated. The active ingredient (purgative principle) is a chemical compound known as Aloin (also called Barbaloin) [19]. The gel has also been found to be rich in antioxidant polyphenols, indoles, and alkaloids. Tests carried out have shown that the nonflavonoid polyphenols contribute to the majority of the total polyphenol content. With this phytochemical profile, A. ferox leaf gel has been identified to be very promising in alleviating symptoms associated with/or prevention of common noncommunicable diseases such as cardiovascular diseases, cancer, neurodegeneration, and diabetes [32].

The leaves have been reported to contain two juices; the yellow bitter sap is used as laxative while the white aloe gel is used in health drinks and skin care products. This purgative drug is used for stomach complaints, mainly as a laxative to "purify" the stomach and also as a bitter tonic (amarum) in various digestives and stomachics (such as

"Lewensessens" and "Swedish Bitters"). Usually, a small crystal of the drug (0.05–0.2 g) is taken orally as a laxative. Half the laxative dose is suggested for arthritis. The fresh bitter sap is instilled directly against conjunctivitis and sinusitis [24, 25]. It is well known that bitter substances stimulate the flow of gastric juices and in so doing improve digestion. The fresh juice emanating from the cut leaf is also applied on burn wounds [33]. A. ferox is claimed to detoxify the damaged surface area and exhibit analgesic and anesthetic properties while promoting new tissue formation (granulation) which fills the wound. It was demonstrated that A. ferox enriched with aloins can inhibit collagenase and metalloprotease activity, which can degrade collagen connective tissues. The effect of A. ferox whole leaf juice on wound healing and skin repair was investigated in an animal model and its safety was evaluated. The results showed that the A. ferox whole leaf juice preparation accelerates wound closure and selectively inhibits microbial growth. No dermal toxicity or side effects were observed during the experimental period [23].

Artemisia herba-alba Asso **(Med)—Asteraceae—Wormwood**

Artemisia herba-alba is commonly known as wormwood or desert wormwood (known in Arabic as shih, and as Armoise blanche in French). It is a greyish strongly aromatic perennial dwarf shrub native to the Northern Africa, Arabian Peninsula, and Western Asia [34]. A. herba-alba has been used in folk medicine by many cultures since ancient times. In Moroccan folk medicine, it is used to treat arterial hypertension and diabetes and in Tunisia, it is used to treat diabetes, bronchitis, diarrhea, hypertension, and neuralgias [34, 35]. Herbal tea from A. herba-alba has been used as analgesic, antibacterial, antispasmodic, and hemostatic agents in folk medicines [34–39]. During an ethnopharmacological survey carried out among the Bedouins of the Negev desert, it was found thatA. herba-alba was used to mitigate stomach disorders. This plant is also suggested to be important as a fodder for sheep and for livestock in the plateau regions of Algeria where it grows abundantly. It has also been reported that Ascaridae from hogs and ground worms were killed by the oil of the Libyan A. herba-alba in a short time [34–42].

Oral administration of 0.39 g/kg body weight of the aqueous extract of the leaves or barks of A. herba-alba has been documented to produce a significant reduction in blood glucose level, while the aqueous extract of roots and methanolic extract of the aerial parts of the plant produce almost no reduction in blood glucose level. The extract of the aerial parts of the plant seems to have minimal adverse effect and high LD_{50} value [19, 30].

Among A. herba-alba phytochemical constituents, essential oils have been extensively studied, with several chemotypes being recognized. The variability from the essential oils isolated from A. herba-alba collected in Algeria, Israel, Morocco, and Spain was revised by Dob and Benabdelkader [43, 44], but, since then, many other studies have reinforced its high chemical polymorphism. Recently, fifty components were identified in A. herba-alba oils, oxygen-containing monoterpenes being dominant in all cases (72–80%). Camphor (17–33%), -thujone (7–28%), and chrysanthenone (4–19%) were the major oil components. Despite the similarity in main components, three types of oils could be defined: (a) -thujone : camphor (23–28 : 17–28%), (b) camphor : chrysanthenone (33 : 12%), and (c) -thujone : camphor : chrysanthenone (24 : 19 : 19%) [43].

The antifungal activity of Artemisia herba-alba was found to be associated with two major volatile compounds isolated from the fresh leaves of the plant. Carvone and piperitone were isolated from Artemisia herba-alba. The antifungal activity of the purified compounds carvone and piperitone was estimated to be 5 μg/mL and 2 μg/mL against Penicillium citrinum and 7 μg/mL and 1.5 μg/mL against Mucor rouxii, respectively. In another study, the antifungal activity of the constituents and biological activities of Artemisia herba-alba essential oils of 25 Moroccan medicinal plants, including A. herba-alba, against Penicillium digitatum, Phytophthora citrophthora, Geotrichum citri-aurantii, and Botrytis cinerea have been reported [42, 43].

Aspalathus linearis **(Brum.F.) R. Dahlg. (Fabaceae)—Rooibos**

Aspalathus linearis, an endemic South African fynbos species, is cultivated to produce the well-known herbal tea, also commonly known as rooibos. Its caffeine-free and comparatively low tannin status, combined with its potential health-promoting properties, most notably antioxidant activity, has contributed to its popularity and consumer acceptance globally. The utilization of rooibos has also moved beyond a herbal tea to intermediate value-added products such as extracts for the beverage, food, nutraceuticals and cosmetic markets [45–52].

Rooibos is used traditionally throughout Africa in numerous ways. It has been used as a refreshment drink and as a healthy tea beverage [8, 19]. It was only after the discovery that an infusion of rooibos, when administered to her colicky baby, cured the chronic restlessness, vomiting, and stomach cramps that rooibos became well known as a "healthy" beverage, leading to a broader consumer base. Many babies since then have been nurtured with rooibos—either added to their milk or given as a weak brew [8, 16, 45–52].

Animal studies have suggested that it has potent antioxidant, immunomodulating, and chemopreventive effects. The plant is rich in minerals and low in tannins. Among the flavonoids present are the unique C-glucoside dihydrochalcones: aspalathin and nothofagin and with aspalathin being the most abundant. In vitro data has shown that the daily intake of the alkaline extracts of the red rooibos tea could suppress HIV infections in the extract, though clinical evaluation has yet to be conducted [8, 45]. There is growing evidence that the flavonoids present in the plant contribute substantially to a reduction in cardiovascular disease and other ailments associated with ageing. Recent studies have shown that aspalathin has beneficial effects on glucose homeostasis in Type 2 diabetes through stimulating glucose uptake in muscle tissues and insulin secretion from pancreatic beta-cells [8, 48]. The unfermented rooibos has been found to have greater chemoprotective effects than the fermented variety [49]. Aspalathin has free-radical capturing properties and is absorbed through the small intestine as such [50].

The bronchodilator, antispasmodic, and blood pressure lowering effects of rooibos tea have been confirmed in vitro and in vivo. It has also been reported that the antispasmodic effect of the rooibos is mediated predominantly through potassium ionchannel activation [51, 52]. There is also increasing evidence of antimutagenic effects. Animal study suggested the prevention of age-related accumulation of lipid peroxidases in the brain [19, 26, 47].

Rooibos is becoming more popular in western countries particularly among health-conscious consumers, due to the absence of alkaloids and low tannin content. It is also reported to have a high level of antioxidants such as aspalathin and nothofagin. The antioxidative effect has also been attributed to the presence of water-soluble polyphenols such as rutin and quercetin [53]. Rooibos is purported to assist nervous tension, allergies, and digestive problems.

Rooibos extracts, usually combined with other ingredients, are available in pill form, but these products fall in the category of dietary supplements. Recent research has underscored the potential of aspalathin and selected rooibos extracts such as an aspalathin-enriched green rooibos extract as antidiabetic agents [54–68]. A patent application for the use of aspalathin in this context was filed in Japan, while a placebo-controlled trial application was filed for the use of rooibos extract as an antidiabetic agent [66]. It is claimed that rooibos extract and a hetero-dimer containing aspalathin, isolated from rooibos, could be used as a drug for the treatment of neurological and psychiatric disorders of the central nervous system [19, 68]. Other opportunities may lie with topical skin products. Two studies concerning inhibition of COX-2 in mouse skin [67] and inhibition of mouse skin tumor promotion tend to support the role of the topical application of rooibos extract. Nonetheless, more research would be needed to explore its potential in preventing skin cancer in humans [67].

Centella asiatica **(L.) Urb. (Apiaceae)—Centella**

Centella asiatica is a medicinal plant that has been used since prehistoric times. It has a pan-tropical distribution and is used in many healing cultures, including Ayurvedic medicine, Chinese traditional medicine, Kampo (Japanese traditional medicine), and African traditional medicine [19, 69]. To date, it continues to be used within the structure of folk medicine and is increasingly being located at the interface between traditional and modern scientifically oriented medicine. Traditionally, C. asiatica is used mainly for wound healing, burns, ulcers, leprosy, tuberculosis, lupus, skin diseases, eye diseases, fever, inflammation, asthma, hypertension, rheumatism, syphilis, epilepsy, diarrhea, and mental illness and is also eaten as a vegetable or used as a spice. In Mauritius, the application of C. asiatica in the treatment of leprosy was reported for the first time in 1852 while the clinical use of C. asiatica, as a therapeutic agent suitable for the treatment of leprous lesions, has been documented since 1887 [19].

The active constituents are characterized by their clinical effects in the treatment of chronic venous disease, wound healing, and cognitive functions amongst others [19]. C. asiatica contains a variety of pentacyclic triterpenoids that have been extensively studied. Asiaticoside and madecassoside are the two most important active compounds that are used in drug preparations. Both are commercially used mainly as wound-healing agent, based on their anti-inflammatory effects. One of the main active constituents of C. asiatica is the ursane-type triterpene saponin, asiaticoside, which is responsible for wound healing properties [19, 70, 71] and is known to stimulate type 1 collagen synthesis in fibroblast cells [72]. Plants collected from various geographical regions and locations in India, Madagascar, Malaysia, Sri Lanka, Andaman Islands, and South Africa have yielded concentrations of asiaticoside ranging from 0.006 to 6.42% of dry weight [73, 74]. C. asiatica also contains several other triterpene saponins. Madecassoside always co-occurs with asiaticoside as a main compound and other saponins have been reported, such as asiaticosides A to G, centelloside, brahmoside, and many others [19, 75]. Madagascar plays a major role in C. asiatica trade. It is the first producer of C. asiatica products worldwide and due to a higher Asiaticoside content of dried leaves, Malagasy origin is appreciated by industry [9].

The ethyl acetate fraction of C. asiatica has been reported to increase the effect of the i.p. administrated antiepileptic drugs phenytoin, valproate, and gabapentin [75, 76] and was found to decrease the pentylenetetrazol- (PTZ-) kindled induced seizures in rats [75, 77]. This effect might be due to an increase in gamma-aminobutyric acid (GABA) levels caused by the extract as reported by Chatterjee et al. [78]. The neuroprotective properties of the plant in monosodium glutamate treated rats were investigated by Ramanathan et al. [79]. The general behavior, locomotor activity, and the CA1 region of the hippocampus were protected by C. asiatica extracts. The levels of catalase, superoxide dismutase, and lipid peroxidase in the hippocampus and striatum were improved indicating a neuroprotective property of the extract [74]. Additionally, the effect of C. asiatica on cognitive function of healthy elderly volunteer was evaluated in a randomized, placebo-controlled, double-blind study involving 28 healthy elderly participants. The subjects have received the plant extract at various doses ranging from 250 to 500 and 750 mg once daily for 2 months, and cognitive performance and mood modulation were assessed. It was found that high dose of the plant extract enhanced working memory and increased N100 component amplitude of event-related potential. Improvements of self-rated mood were also found following the C. asiatica treatment. The high dose of C. asiatica used in the study was suggested to increase calmness and alertness after 1 and 2 months of treatment. However, the precise mechanism(s) underlying these effects still require further investigation [72].

Catharanthus roseus (L.) G. Don (Apocynaceae)—Madagascan Periwinkle

Catharanthus roseus (Madagascar periwinkle) is a well-known medicinal plant that has its root from the African continent. The interest in this species arises from its therapeutic role, as it is the source of the anticancer alkaloids vincristine and vinblastine, whose complexity renders them impossible to be synthesized in the laboratory; the leaves of this species are still, today, the only source [8, 12, 19, 80]. C. roseus originates from Madagascar but now has a wide distribution throughout the tropics, and the story on the traditional utilisation of this plant can be retraced to Madagascar where healers have been using it extensively to treat panoply of ailments. It is commonly used in traditional medicine as a bitter tonic, galactogogue, and emetic. Application for treatment of rheumatism, skin disorders, and venereal diseases has also been reported [8,10, 19].

C. roseus has been found to contain a plethora of phytochemicals (as many as 130 constituents) and the principal component is vindoline (up to 0.5%). Other biologically active compounds are serpentine, catharanthine, ajmalicine (raubasine), akuammine, lochnerine, lochnericine, and tetrahydroalstonine. The plant is also rich in bisindole alkaloids; vindoline and catharanthine are found in very small amounts: vincristine (=leurocristine) in up to 3 g/t of dried drug and vinblastine (=vincaleucoblastine) in a slightly larger amount [8, 12, 19,81–84].

The oral administration of water-soluble fractions and ethanolic extracts of the leaves have been found to show significant dose-dependent reduction in the blood sugar at 4 h by 26.2, 31.4, 35.6, and 33.4%, respectively, in normal rats. In addition, oral administration of 500 mg/kg 3.5 h before oral glucose tolerance test (10 mg/kg) and 72 h after STZ administration (50 mg/kg IP) in rats showed significant antihyperglycaemic effects. No gross behavioural changes and toxic effects were observed up to 4 mg/kg IP [85].

Extract at dose of 500 mg/kg given orally for 7 and 15 days showed 48.6 and 57.6% hypoglycemic activity, respectively. Prior treatment at the same dose for 30 days provided complete protection against streptozotocin (STZ) challenge (75 mg/kg/i.p. × 1). Enzymatic activities of glycogen synthase, glucose 6-phosphate-dehydrogenase, succinate dehydrogenase, and malate dehydrogenase were decreased in liver of diabetic animals in comparison to normal ones and were significantly improved after treatment with extract at dose of 500 mg/kg p.o. for 7 days. Results indicate increased metabolization of glucose in treated rats. Increased levels of lipid peroxidation measured as 2-thiobarbituric acid reactive substances indicative of oxidative stress in diabetic rats were also normalized by treatment with the extract [2].

Vincristine and Vinblastine are antimitotics as they bind to tubulin and prevent the formation of microtubules that assist in the formation of the mitotic spindle; in this way, they block mitosis in the metaphase. These alkaloids are highly toxic; they both have neurotoxic activity (especially vincristine) because the microtubule assembly also plays a role in neurotransmission. Their peripheral neurotoxic effects are neuralgia, myalgia, paresthesia, loss of the tendon reflexes, depression, and headache, and their central neurotoxic effects are convulsive episodes and respiratory difficulties. Other side effects are multiple and include alopecia, gastrointestinal distress including constipation, ulcerations of the mouth, amenorrhoea, and azoospermia [10, 19, 80, 84, 85].

Recently, new phenolics in different plant parts (leaves, stems, seeds, and petals), including flavonoids and phenolic acids, were reported [10, 19, 82]. In addition, a phytochemical study concerning several classes of metabolites was performed and bioactivity was assessed [81–84]. The high antioxidant potential of C. roseus was demonstrated in vitro using the radicals DPPH, superoxide, and nitric oxide [19, 81].

Cyclopia genistoides (L.) Vent. (Fabaceae)—Honeybush

Cyclopia genistoides is an indigenous herbal tea to South Africa and considered as a health food. Traditionally, the leafy shoots and flowers were fermented and dried to prepare tea. It has also been used since early times for its direct positive effects on the urinary system and is valued as a stomachic that aids weak digestion without affecting the heart. It is a drink that is mainly used as a tea substitute because it contains no harmful substances such as caffeine. It is one of the few indigenous South African plants that made the transition from the wild to a commercial product during the past 100 years. Research activities during the past 20 years have been geared towards propagation, production, genetic improvement, processing, composition, and the potential for value adding [19, 25–31, 45].

A decoction of honeybush was used as a restorative and as an expectorant in chronic catarrh and pulmonary tuberculosis [25–31, 86]. Drinking an infusion of honeybush apparently also increases the appetite, but no indication is given of the specific species [87]. According to Marloth [88], honeybush was praised by many colonists as being wholesome, valuing it as a stomachic that aids weak digestion without producing any serious stimulating effects on the heart. It also alleviates heartburn and nausea [25–31]. Anecdotal evidence suggests that it stimulates milk production in breast-feeding women and treats colic in babies [89].

Modern use of honeybush is prepared as an infusion and at times taken together with rooibos. The tea bags usually contain a larger percentage of rooibos than honeybush. Parts of other indigenous South African plants and fruits mixed with honeybush include dried buchu leaves, pieces of African potato (Hypoxis hemerocallidea) corms, and dried marula (Sclerocarya birrea) fruit. The ready-to-drink honeybush iced tea market is not developed to the same extent as that of rooibos, while honeybush "espresso," amongst others, has not been tried [10, 19].

Honeybush is well known as caffeine-free, low tannin, aromatic herbal tea with a wealth of polyphenolic compounds associated with its health-promoting properties. The exact biologically active phytochemicals from honeybush are unknown, but the beneficial effects have been justified based on phenolic compounds. The major compounds, present in all species analyzed to date, are the xanthones, mangiferin and isomangiferin, and the flavanone, hesperidin. Recently, two benzophenone derivatives 3-C-β-glucosides of maclurin and iriflophenone were isolated for the first time from C. genistoides and were tested for pro-apoptotic activity toward synovial fibroblasts isolated from rheumatoid arthritis patients. The strongest proapoptotic activity was obtained for isomangiferin and iriflophenone 3-C-β-glucoside, which were not previously evaluated as potential antiarthritic agents. Proapoptotic effects were recorded for mangiferin and hesperidin, which are major polyphenols in all commercially used honeybush plants. Recently, C. genistoides has attracted much attention for the production of an antioxidant product high in mangiferin content. The latter and its sustainability make C. genistoides an attractive source of mangiferin. Other potential applications are for the prevention of skin cancer, alleviation of menopausal symptoms, and lowering of blood glucose levels [10, 19, 45, 90].

Harpagophytum procumbens **(Burch.) DC. (Pedaliaceae)—Devil's Claw**

Harpagophytum procumbens is native to the red sand areas in the Transvaal of South Africa, Botswana, and Namibia. It has spread throughout the Kalahari and Savannah desert regions. The indigenous San and Khoi peoples of Southern Africa have used Devil's Claw medicinally for centuries, if not millennia [8, 19, 91]. Harpagophytum procumbens has an ancient history of multiple indigenous uses and is one of the most highly commercialized indigenous traditional medicines from Africa, with bulk exports mainly to Europe where it is made into a large number of health products such as teas, tablets, capsules, and topical gels and patches [92].

Traditional uses recorded include allergies, analgesia, anorexia, antiarrhythmic, antidiabetic, antiphlogistic, antipyretic, appetite stimulant, arteriosclerosis, bitter tonic, blood diseases, boils (topical), childbirth difficulties, choleretic, diuretic, climacteric (change of life) problems, dysmenorrhea, dyspepsia, edema, fever, fibromyalgia, fibrositis, gastrointestinal disorders, headache, heartburn, indigestion, liver and gall bladder tonic, malaria, migraines, myalgia, neuralgia, nicotine poisoning, sedative, skin cancer (topical), skin ulcers (topical), sores (topical), tendonitis, urinary tract infections, and vulnerary for skin injuries. The major clinical uses for Devil's claw are as an anti-inflammatory and analgesic in joint diseases, back pain, and headache. Evidence from scientific studies in animals and humans has resulted in widespread use of standardized Devil's claw as a mild analgesic for joint pain in Europe [8, 12, 19, 91–95].

Chemical constituents according to the published literature include potentially (co)active chemical constituents: iridoid glycosides (2.2% total weight) harpagoside (0.5–1.6%), 8-p-coumaroylharpagide, 8-feruloyl-harpagide, 8-cinnamoylmyoporoside, pagoside, acteoside, isoacteoside, 6′-O-acetylacteoside, 2,6-diacetylacteoside, cinnamic acid, caffeic acid, procumbide, and procumboside. The constituent 6-acetylacteoside, being present in H. procumbens and absent in H. zeyheri, allows users to distinguish between the two species. Other compounds include flavonoids, fatty acids, aromatic acids, harpagoquinone, stigmasterol, beta-sitosterol, triterpenes, sugars (over 50%), and gum resins [8, 12, 19, 91–95]. Harpagoside isolated from H. procumbensvaries within the plant and is the highest in the secondary tubers, with lower levels in the primary roots. The flowers, stems, and leaves appear to be devoid of active compounds. Iridoid glycosides isolated from H. procumbens tend to show dose-dependent anti-inflammatory and analgesic effects equivalent to phenylbutazone; they are apparently inactivated by gastric acids. Harpagoside is the most effective when given parenterally and loses potency markedly when given by mouth; enteric coated preparations might maintain efficacy despite exposure to gastric acids. Harpagoside has also been reported to inhibit arachidonic acid metabolism through both cyclooxygenase and lipoxygenase pathways [9, 91–96].

There are varying and contradictory opinions regarding the anti-inflammatory and analgesic effects of Devil's claw in the treatment of arthritic diseases and low back pain. The controversy revolves around the active constituent of Devil's claw and its mechanism of action, as it appears to be different than that of nonsteroidal anti-inflammatory drugs (NSAID). The evidence from scientific studies in animals and humans has resulted in widespread use of standardized Devil's claw as a mild analgesic for joint pain in Europe [8, 12, 19, 91–96].

Several clinical studies have been performed to determine the effectiveness ofH. procumbens for its use as anti-inflammatory, general analgesic (commonly for lower back pain), and anti-rheumatic agent. To determine the effectiveness on lower back pain, Harpagophytum extract WS1351 was administered in two daily doses of 600 and 1200 mg containing 50 and 100 mg of harpagoside, respectively, and compared to placebo. This randomized double-blind study took place over 4 weeks and subjects (n=197) with chronic susceptibility to back pain and current exacerbations with intense pain were included. Out of 183 subjects that completed the trial, six in the 600 mg and 10 in the 1200 mg were reported "pain-free" without using Tramadol (rescue pain medication). However, data analyses suggested that the 600 mg group reaped more benefits where less severe pain and no radiation or neurological deficit was present. The patients with more severe pain tended to use more Tramadol but not to the maximum permitted dose [92].

Momordica charantia Linn. (Cucurbitaceae)—Bitter Melon

Momordica charantia, also known as bitter melon, is a tropical vegetable grown throughout Africa. The leaf may be made into a tea called "cerassie," and the juice, extracted from the various plant parts (fruit pulp, seeds, leaves, and whole plant), is very common folklore remedy for diabetes [2]. M. charantia has a long history of use as a folklore hypoglycemic agent where the plant extract has been referred to as vegetable insulin [97].

Several active compounds have been isolated from M. charantia, and some mechanistic studies have been done [98–101]. Khanna et al. [102] have reported the isolation from fruits, seeds, and tissue culture of seedlings, of "polypeptide-p," a 17-amino acid, 166-residue polypeptide which did not cross-react in an immunoassay for bovine insulin. Galactose binding lectin with a molecular weight of 124,000 isolated from the seeds of M. charantia has been evaluated for its antilipolytic and lipogenic activities in isolated rat adipocytes and found comparable to insulin [103, 104]. Extracts of fruit pulp, seed, leaves, and whole plant of M. charantia have shown hypoglycemic effect in various animal models [103, 104]. Karunanayake et al.

[105] found a significant improvement in glucose tolerance and hyperglycemia when the fruit juice of bitter melon was administered orally to rats. Fresh as well as freeze-dried M. charantia was found to improve glucose tolerance significantly in normal rats and noninsulin dependent diabetics (NIDDM) [105–108]. It was hypothesized that M. charantia fruit contains more than one type of hypoglycemic component. These may include an active principle called "charantin," a homogenous mixture of β-sitosterol-glucoside and 2-5-stigmatadien-3-β-ol-glucoside that can produce a hypoglycemic effect in normal rabbits [109]. Shibib et al. [110] in a study administered the ethanolic extract of the fruit of bitter melon to STZ diabetic rats orally at a dose equivalent to 200 mg extract per kg body weight. Ninety minutes after the administration, they found that blood glucose levels had been reduced by 22%. The glucogenic enzymes-glucose-6-phosphatase and fructose 1, 6-bisphosphatase in the liver were also depressed. Further evidence for a beneficial chronic effect is that an improvement in both glucose tolerance and fasting blood glucose levels was observed in 8 NIDDM patients following 7 weeks of daily consumption of powered M. charantia fruit [97]. Although some authors have indicated that the effect of M. charantia is not associated with an increase in circulating insulin, Welihinda et al. [111] and Welihinda and Karunanayake [112] demonstrated that an aqueous extract from the fruit of M. charantia was a potent stimulator of insulin release from β-cell-rich pancreatic islets isolated from obese-hyperglycaemic mice. Recently, Matsuura et al. [113] have isolated an α-glucosidase inhibitor from M. charantia seeds which can be a potential drug therapy for postprandial hyperglycaemia (PPHG). However, Dubey et al. [104] found that the aqueous, methanolic, and saline extracts of M. charantia produced a significant hypoglycaemic effect in rats. In addition, the methanol and saline extracts also prevented adrenaline-induced hyperglycaemia.

When tested on laboratory animals, bitter melon has shown hypoglycaemic as well as antihyperglycaemic activities. Polypeptide-p isolated from fruit, seeds, and tissue of bitter melon showed potent hypoglycaemic effects when administered subcutaneously to gerbils, langurs, and humans. The aqueous extracts of M. charantia improved oral glucose tolerance test (OGTT) after 8 h in normal mice and reduced hyperglycaemia by 50% after 5 h in STZ diabetic mice. In addition, chronic oral administration of extract to normal mice for 13 days improved OGTT while no significant effect was seen on plasma insulin levels. We recently reported that M. charantia fruit extracts have a direct impact on transport of glucose in vitro [99, 114–117].

Pelargonium sidoides DC. (Geraniaceae)—Umckaloabo

Pelargonium sidoides is native to the coastal regions of South Africa, and available ethnobotanical information shows that the tuberous P. sidoides is an important traditional medicine with a rich ethnobotanical history [19].

P. sidoides root extract EPs 7630, also known as Umckaloabo, is a herbal remedy thought to be effective in the treatment of acute respiratory infections. There are numerous studies about P. sidoides and respiratory tract infections [118, 119]. These studies showed that P. sidoides may be effective in alleviating symptoms of acute rhinosinusitis and the common cold in adults, but doubt exists. It may be effective in relieving symptoms of acute bronchitis in adults and children and sinusitis in adults [118]. EPs 7630 significantly reduced bronchitis symptom scores in patients with acute bronchitis by day 7 [119]. No serious adverse events were reported. EPs 7630 has a positive effect on phagocytosis, oxidative burst, and intracellular killing of cells [120–125]. P. sidoides extract modulates the production of secretory immunoglobulin A in saliva, both interleukin-15 and interleukin-6 in serum, and interleukin-15 in the nasal mucosa [19, 126].

In one research, P. sidoides was documented to represent an effective treatment against common cold. It was reported to significantly reduce the severity of symptoms and shortens the duration of the common cold compared with placebo. Because of these effects, the authors have aimed at establishing whether or not P. sidoides could affect the asthma attack frequency after upper respiratory tract viral infection. Results for some 20 clinical trials have been published, 7 of which were observational studies and the remaining 13 were randomized, double-blind, and placebo-controlled. For 6 of those 13 trials, only preliminary results have been published. All trials have been carried out using EPs 7630 in liquid or solid forms. The data derived from these trials has been evaluated in 2 reviews [19, 118, 119].

Antibacterial activity of extracts and isolated constituents of P. sidoides was evaluated by Kayser [127] against three gram-positive and five gram-negative bacteria. Most compounds exhibited antibacterial activities. Further investigations by Lewu et al. [128] have supported these findings. EPs 7630 has been found to show a synergistic indirect antibacterial effect in group A-streptococci (GAS) through inhibition of bacterial adhesion to human epithelial cells as well as induction of bacterial adhesion to buccal epithelial cells [8, 129]. Wittschier et al. [130] and Beil and Kilian [131] confirmed the antiadhesive effect of EPs 7630 for Helicobacter pylori growth and adhesion to gastric epithelial cells. Umckaloabo has been documented to significantly stimulate phagocytosis, oxidative burst, and intracellular killing whichwas also enhanced [120–125]. It was proposed that modulation of epithelial-cell bacteria interaction through EPs 7630 may protect mucous membranes from microorganisms evading host defense mechanisms. These findings tend to provide a rationale for the treatment of upper respiratory tract infections with EPs 7630 [131, 132].

Taylor et al. [133] have recently established the antimycobacterial activity for hexane extracts of roots of P. reniforme and P. sidoides. Gödecke et al. [134] have reported no significant effect on the bacterial growth of two strains of mycobacteria by extracts and fractions of P. sidoides. An antitubercular effect may thus be achieved indirectly by stimulation of immune response. This assumption was supported by Mativandlela et al. [135] as none of the compounds isolated from P. sidoides showed any significant activity against M. tuberculosis.

Kayser et al. [136] have investigated into extracts and isolated constituents of P. sidoides for their effects on nonspecific immune functions in various bioassays. They found indirect activity, possibly through activation of macrophage functions. Activation was confirmed through the presence of tumor necrosis factor (TNF-alpha) and inorganic nitric oxides (iNO). Kolodziej [137, 138] also observed TNF-inducing potencies for EPs 7630 as well as interferon-like activities. Koch et al. [139] observed interferon- (IFN-) beta production increased and natural killer cell mediated cytotoxicity enhanced in MG-63 human osteosarcoma cells preincubated with Umckaloabo [19, 139]. Kolodziej [137, 138] investigated polyphenol-containing extracts of P. sidoides and simple phenols, flavan-3-ols, proanthocyanidins, and hydrolysable tannins for gene expressions (iNOS, IL-1, IL-10, IL-12, IL-18, TNF-alpha, and IFN-alpha/gamma). All extracts and compounds were capable of enhancing the iNOS and cytokine mRNA levels in parasitised cells.

DISCUSSION

Medicinal plants are an integral part of the African healthcare system since time immemorial. Interest in traditional medicine can be explained by the fact that it is a fundamental part of the culture of the people who use it and also due to the economic challenge: on one side, the pharmaceutical drugs are not accessible to the poor and on the other side, the richness and diversity of the fauna and flora of Africa are an inexhaustible source of therapies for panoply of ailments [140]. Nonetheless, there is still a paucity of clinical evidence to show that they are effective and safe for humans. Without this information, users of traditional medicinal plants in Africa and elsewhere remain skeptical about the value of such therapies. This denies people the freedom to choose plants that are potentially less costly and are more accessible. Another issue concerning the use of botanical remedies is the need to understand the safety of these therapies. For these reasons, information about efficacy and safety of traditional medicines is urgently required. The present paper has endeavoured to overview just a few common medicinal plants from the African continent which have short- as well as long-term potential to be developed as future phytopharmaceuticals to treat and/or manage panoply of infectious and chronic conditions. Within the framework of enhancing the significance of traditional African medicinal plants, aspects such as traditional use, phytochemical composition, and in vitro, in vivo, and clinical studies pertaining to the use of these plants have been explored.

During the last few decades, it has become evident that there exists a plethora of plants with medicinal potential and it is increasingly being accepted that the African traditional medicinal plants might offer potential template molecules in the drug discovery process. Many of the plants presented here show very promising medicinal properties thus warranting further clinical investigations. Nonetheless, only few of them have robust scientific and clinical proofs and with a significant niche market (e.g., Aloe ferox, Artemisia afra, Aspalathus linearis, Centella asiatica, and Pelargonium sidoides) and a lot more have yet to be explored and proved before reaching the global market [7, 8, 12, 19].

In the light of modern science, significant efforts should be geared to identify and characterize the bioactive constituents from these plants. Indeed, the discovery of therapeutic compounds from traditional medicinal plant remedies remains a medically and potentially challenging task. For adventure in such an attempt, highly reproducible and robust innovative bioassays are needed in view of our limited understanding of the multifactorial pathogenicity of diseases. Innovative strategies to improve the process of plant collection are needed, especially with the legal and political issues surrounding benefit-sharing agreements. Since drug discovery from medicinal plants has traditionally been so time-consuming, it is also of uttermost importance for investigators to embark and devise new automated bioassays with special emphasis on high throughput procedures that can screen, isolate, and process data from an array of phytochemicals within shorter time lapse for product development. Additionally, these procedures should also attempt to rule out false positive hits and dereplication methods to remove nuisance compounds [7].

Nonetheless, despite continuous comprehensive and mechanism-orientated evaluation of medicinal plants from the African flora, there is still a dearth of literature coming from the last decade's investigations addressing procedures to be adopted for quality assurance, authentication, and standardization of crude plant products. Appropriate standardization could be achieved via proper management of raw material, extraction procedures, and final product formulation. Without effective quality control, consistency and market value of the herbal product may be compromised. Indeed, one of the main constraints to the growth of a modern African phytomedicine industry has also been identified as the lack of proper validation of traditional knowledge and also the lack of technical specifications and quality control standards. This makes it extremely difficult for buyers, whether national or international, to evaluate the safety and efficacy of plants and extracts, or compare batches of products from different places or from year to year. This is in marked contrast with Europe and Asia where traditional methods and formulations have been recorded and evaluated both at the local and national levels. This would also tend to justify why the level of trade of phytomedicines in Asia and Europe is blooming more than those in Africa [7].

It is also imperative that potential risk factors, for example, the contamination of medicinal plant products with heavy metals from African traditional medicine products, be addressed and that regulatory guidelines are not only carefully developed but also enforced.

Controlled growth (under GACP) and processing environments (under Good Manufacturing Practice) need to ensure that contamination of medicinal plant material is kept to a minimum. For the medicinal plant industry, cultivated plant material is preferred as it is easier to control the supply chain plus contamination is nominal [141]. On the other hand, proper identification of a medicinal plant material is fundamental to the quality control process; it must be established unmistakably that the source of the plant material is genuine. Following this, microbial contamination (fungal, bacterial, and any potential human pathogens) must be checked during the stages of processing of the material. Chemical, pharmacological, and toxicological evaluations, conducted according to the principles of Good Laboratory Practices (GLPs), will certify the bioactive properties of the material undergoing processing. These tests also are often the predictors of safety of the products manufactured. Clinical safety and efficacy will need to be established through exhaustive and usually lengthy trials during the early stages of the development of a therapeutic agent. After that, so long as the standard operating procedures are adhered to, then the unit dosage forms produced will be considered safe. Notwithstanding this, quality assurance procedures must be instituted so that the products coming from the factory are of good quality, safety, and efficacy [142]. To this effect, during the development stage, product standardization, quality control and assurance, double-blind, placebo-controlled, and randomized clinical controlled trials using standardized products or products containing pure plant extracts are essential components that need to be perfected in order to translate the potential of African botanicals into a reality for human to benefit [7, 142].

CONCLUSIONS

It is evident from the literature that there is currently a renewed interest in African-plant-based medicines in the prevention and cure of various pathologies. Medicinal plants still play an important role in healthcare system in African countries. Nonetheless, there are still many major challenges that need to be overcome and addressed for its full potential to be realized as the effective treatment of diseases with plant products has not been validated thoroughly with robust scientific criteria to compete with existing conventional therapies. Additionally, other issues that need to be addressed are that of access and benefit sharing following the Nagoya agreement. Local laws need to be TRIPS compliant if trade of African herbal products is to increase, and, at the same time, issues of sustainable use and development of plant products need to be addressed.

REFERENCES

1. WHO, Fact sheet N°134, 2008,http://www.who.int/mediacentre/factsheets/2003/fs134/en/.
2. A. Gurib-Fakim, "Medicinal plants: traditions of yesterday and drugs of tomorrow," Molecular Aspects of Medicine, vol. 27, no. 1, pp. 1–93, 2006.
3. V. Chintamunnee and M. F. Mahomoodally, "Herbal medicine commonly used against infectious diseases in the tropical island of Mauritius," Journal of Herbal Medicine, vol. 2, pp. 113–125, 2012.
4. H. Nunkoo and M. F. Mahomoodally, "Ethnopharmacological survey of native remedies commonly used against infectious diseases in the tropical island of Mauritius," Journal of Ethnopharmacology, vol. 143, no. 2, pp. 548–564, 2012.
5. S. Shohawon and M. F. Mahomoodally, "Complementary and alternative medicine use among Mauritian women," Complementary Therapies in Clinical Practice, vol. 19, no. 1, pp. 36–43, 2013.
6. Aone Mokaila, 2001, http://www.blackherbals.com/atcNewsletter913.pdf.
7. A. Gurib-Fakim and M. F. Mahomoodally, "African flora as potential sources of medicinal plants: towards the chemotherapy of major parasitic and other infectious diseases- a review," Jordan Journal of Biological Sciences, vol. 6, pp. 77–84, 2013.
8. A. Gurib-Fakim, T. Brendler, L. D. Phillips, and L. N. Eloff, Green Gold—Success Stories Using Southern African Plant Species, AAMPS Publishing, Mauritius, 2010.
9. C. Manach, A. Scalbert, C. Morand, C. Rémésy, and L. Jiménez, "Polyphenols: food sources and bioavailability," American Journal of Clinical Nutrition, vol. 79, no. 5, pp. 727–747, 2004.

10. B. M. Abegaz, B. T. Ngadjui, G. N. Folefoc et al., "Prenylated flavonoids, monoterpenoid furanocoumarins and other constituents from the twigs of Dorstenia elliptica (Moraceae)," Phytochemistry, vol. 65, no. 2, pp. 221–226, 2004.
11. N. R. Farnsworth, O. Akerele, A. S. Bingel, D. D. Soejarto, and Z. Guo, "Medicinal plants in therapy," Bulletin of the World Health Organization, vol. 63, no. 6, pp. 965–981, 1985.
12. Association of African Medicinal Plants Standards (AAMPS), http://www.aamps.org/.
13. S. E. Atawodi, "Antioxidant potential of African medicinal plants,"African Journal of Biotechnology, vol. 4, no. 2, pp. 128–133, 2005.
14. S. O. Okoro, A. H. Kawo, and A. H. Arzai, "Phytochemical screening, antibacterial and toxicological activities of Acacia senegal extracts,"Bayero Journal of Pure and Applied Sciences, vol. 5, no. 1, pp. 163–1170, 2011.
15. R. Jain, P. Sharma, T. Bhagchandani, and S. C. Jain, "Phytochemical investigation and antimicrobial activity of Acacia senegal root heartwood," Journal Pharmaceutical Research, vol. 5, pp. 4934–4938, 2012.
16. A. Gurib-Fakim and M. J. Kasilo, Promoting African Medicinal Plants through an African Herbal Pharmacopoeia. Special Issue 14: Decade of African Traditional Medicine, 2001–2010.
17. B. S. Aliyu, Common Ethnomedicinal Plants of the Semiarid Regions of West Africa, Triumph Publishing, Kano, Nigeria, 2006.
18. S. O. Okoro, A. H. Kawo, and A. H. Arzai, "Phytochemical screening, antibacterial and toxicological activities of Acacia senegal extracts,"Bayero Journal of Pure and Applied Sciences, vol. 5, no. 1, pp. 163–170, 2011.
19. T. Brendler, L. N. Eloff, A. Gurib-Fakim, and L. D. Phillips, African Herbal Pharmacopeia, AAMPS Publishing, Mauritius, 2010.
20. B.-E. van Wyk, "A broad review of commercially important southern African medicinal plants," Journal of Ethnopharmacology, vol. 119, no. 3, pp. 342–355, 2008.
21. B. E. Van Wyk and M. Wink, Medicinal Plants of the World: An Illustrated Scientific Guide to important Medicinal Plants and Their Uses, Briza Publications, Pretoria, South Africa, 2004.

22. Y. Jia, G. Zhao, and J. Jia, "Preliminary evaluation: the effects of Aloe ferox Miller and Aloe arborescens Miller on wound healing," Journal of Ethnopharmacology, vol. 120, no. 2, pp. 181–189, 2008.
23. W. Chen, B.-E. Van Wyk, I. Vermaak, and A. M. Viljoen, "Cape aloes—a review of the phytochemistry, pharmacology and commercialisation ofAloe ferox," Phytochemistry Letters, vol. 5, no. 1, pp. 1–12, 2012.
24. A. Melin, A bitter pill to swallow: a case study of the trade & harvest of Aloe ferox in the Eastern Cape, South Africa [M.S. thesis], Imperial College, 2009.
25. B. E. Van Wyk and N. Gericke, People's Plants: A Guide to Useful Plants of Southern Africa, Briza Publications, Pretoria, South Africa, 2000.
26. B.-E. van Wyk, "A broad review of commercially important southernAfricanmedicinalplants,"JournalofEthnopharmacology, vol. 119, no. 3, pp. 342–355, 2008.
27. B. E. Van Wyk and G. F. Smith, Guide to the Aloes of South Africa, Briza Publications, Pretoria, South Africa, 1996.
28. B. E. Van Wyk and G. F. Smith, Guide to the Aloes of South Africa, Briza Publications, Pretoria, South Africa, 2004.
29. B. E. Van Wyk, B. Van Oudtshoorn, and N. Gericke, Medicinal Plants of South Africa, Briza Publications, Pretoria, South Africa, 2nd edition, 2009.
30. B.-E. Van Wijk, M. C. B. Van Rheede Van Oudtshoorn, and G. F. Smith, "Geographical variation in the major compounds of Aloe ferox leaf exudate," Planta Medica, vol. 61, no. 3, pp. 250–253, 1995.
31. B. E. Van Wyk, A. Yenesew, and E. Dagne, "Chemotaxonomic survey of anthraquinones and pre-anthraquinones in roots of Aloe species,"Biochemical Systematics and Ecology, vol. 23, no. 3, pp. 267–275, 1995.
32. T. L. Du, F. H. Van Der Westhuizen, and L. Botes, "Aloe ferox leaf gel phytochemical content, antioxidant capacity, and possible health benefits," Journal of Agricultural and Food Chemistry, vol. 55, no. 17, pp. 6891–6896, 2007.

33. F. A. Andersen, "Final report on the safety assessment of Aloe andongensis extract, Aloe andongensis leaf juice, Aloe arborescens leaf extract, Aloe arborescens leaf juice, Aloe arborescens leaf protoplasts, Aloe barbadensis flower extract, Aloe barbadensis leaf, Aloe barbadensis leaf extract, Aloe barbadensis leaf juice, Aloe barbadensis leaf polysaccharides,Aloe barbadensis leaf water, Aloe ferox ferox leaf extract," International Journal of Toxicology, vol. 26, no. 2, pp. 1–50, 2007.
34. R. Segal, I. Feuerstein, and A. Danin, "Chemotypes of Artemisia herba-alba in Israel based on their sesquiterpene lactone and essential oil constitution," Biochemical Systematics and Ecology, vol. 15, no. 4, pp. 411–416, 1987.
35. A. Ziyyat, A. Legssyer, H. Mekhfi, A. Dassouli, M. Serhrouchni, and W. Benjelloun, "Phytotherapy of hypertension and diabetes in oriental Morocco," Journal of Ethnopharmacology, vol. 58, no. 1, pp. 45–54, 1997.
36. A. Tahraoui, J. El-Hilaly, Z. H. Israili, and B. Lyoussi, "Ethnopharmacological survey of plants used in the traditional treatment of hypertension and diabetes in south-eastern Morocco (Errachidia province)," Journal of Ethnopharmacology, vol. 110, no. 1, pp. 105–117, 2007.
37. N.-A. Zeggwagh, O. Farid, J. B. Michel, and M. Eddouks, "Cardiovascular effect of Artemisia herba alba aqueous extract in spontaneously hypertensive rats," Methods and Findings in Experimental and Clinical Pharmacology, vol. 30, no. 5, pp. 375–381, 2008.
38. M. Laid, M.-E. F. Hegazy, A. A. Ahmed, K. Ali, D. Belkacemi, and S. Ohta, "Sesquiterpene lactones from Algerian Artemisia herba-alba,"Phytochemistry Letters, vol. 1, no. 2, pp. 85–88, 2008.
39. J. Friedman, Z. Yaniv, A. Dafni, and D. Palewitch, "A preliminary classification of the healing potential of medicinal plants, based on a rational analysis of an ethnopharmacological field survey among Bedouins in the Negev Desert, Israel," Journal of Ethnopharmacology, vol. 16, no. 2-3, pp. 275–287, 1986.
40. F. Fenardji, M. Klur, C. Fourlon, and R. Ferrando, "White Artemisia(Artemisia herba alba L.)," Revue d"elevage et de medecine veterinaire des pays tropicaux, vol. 27, no. 2, pp. 203–206, 1974.

41. A. Benmansur, S. A. Taleb-Bendiab, N. Mashev, and G. Vasilev, "Studies on the chemical composition of Artemisia (Artemisia herba-alba),"Bolgarskoi Akademii Nauk, vol. 43, no. 8, pp. 65–67, 1990.
42. A. E.-H. H. Mohamed, M. A. El-Sayed, M. E. Hegazy, S. E. Helaly, A. M. Esmail, and N. S. Mohamed, "Chemical constituents and biological activities of Artemisia herba-alba," Records of Natural Products, vol. 4, no. 1, pp. 1–25, 2010.
43. R. Belhattab, L. Amor, J. G. Barroso, L. G. Pedro, and A. C. Figueiredo, "Essential oil from Artemisia herba-albaAsso grown wild in Algeria: variability assessment and comparison with an updated literature survey," Arabian Journal of Chemistry, vol. 57, no. 4, pp. 603–619, 2012.
44. T. Dob and T. Benabdelkader, "Chemical composition of the essential oil of Artemisia herba-albaAsso grown in Algeria," Journal of Essential Oil Research, vol. 18, no. 6, pp. 685–690, 2006.
45. E. Joubert and D. de Beer, "Rooibos (Aspalathus linearis) beyond the farm gate: from herbal tea to potential phytopharmaceutical," South African Journal of Botany, vol. 77, no. 4, pp. 869–886, 2011.
46. F. R. Van Heerden, B.-E. Van Wyk, A. M. Viljoen, and P. A. Steenkamp, "Phenolic variation in wild populations of Aspalathus linearis (rooibos tea)," Biochemical Systematics and Ecology, vol. 31, no. 8, pp. 885–895, 2003.
47. B. E. Van Wyk and G. H. Verdoorn, "Alkaloids of the genera Aspalathus, Rafnia and Wiborgia (Fabaceae-Crotalarieae)," South African Journal of Botany, vol. 55, pp. 520–522, 1989.
48. A. Kawano, H. Nakamura, S.-I. Hata, M. Minakawa, Y. Miura, and K. Yagasaki, "Hypoglycemic effect of aspalathin, a rooibos tea component from Aspalathus linearis, in type 2 diabetic model db/db mice,"Phytomedicine, vol. 16, no. 5, pp. 437–443, 2009.
49. J. L. Marnewick, F. H. van der Westhuizen, E. Joubert, S. Swanevelder, P. Swart, andW. C. A. Gelderblom, "Chemoprotective properties of rooibos (Aspalathus linearis), honeybush (Cyclopia intermedia) herbal and green and black (Camellia sinensis) teas against cancer promotion induced by fumonisin B1 in rat liver," Food and Chemical Toxicology, vol. 47, no. 1, pp. 220–229, 2009.

50. S. Kreuz, E. Joubert, K.-H. Waldmann, and W. Ternes, "Aspalathin, a flavonoid in Aspalathus linearis (rooibos), is absorbed by pig intestine as a C-glycoside," Nutrition Research, vol. 28, no. 10, pp. 690–701, 2008.

51. A. H. Gilani, A.-U. Khan, M. N. Ghayur, S. F. Ali, and J. W. Herzig, "Antispasmodic effects of Rooibos tea (Aspalathus linearis) is mediated predominantly through K^+-channel activation," Basic and Clinical Pharmacology and Toxicology, vol. 99, no. 5, pp. 365–373, 2006.

52. A.-U. Khan and A. H. Gilani, "Selective bronchodilatory effect of Rooibos tea (Aspalathus linearis) and its flavonoid, chrysoeriol,"European Journal of Nutrition, vol. 45, no. 8, pp. 463–469, 2006.

53. R. Fukasawa, A. Kanda, and S. Hara, "Anti-oxidative effects of rooibos tea extract on autoxidation and thermal oxidation of lipids," Journal of Oleo Science, vol. 58, no. 6, pp. 275–283, 2009.

54. E. Joubert, "Effect of batch extraction conditions on yield of soluble solids from rooibos tea," International Journal of Food Science and Technology, vol. 23, pp. 43–47, 1988.

55. E. Joubert, "Chemical and sensory analyses of spray- and freeze-dried extracts of rooibos tea (Aspalathus linearis)," International Journal of Food Science and Technology, vol. 25, pp. 344–349, 1990.

56. E. Joubert, "Effect of batch extraction conditions on yield of polyphenols from rooibos tea (Aspalathus linearis)," International Journal of Food Science and Technology, vol. 25, pp. 339–343, 1990.

57. E. Joubert, "Tristimulus colour measurement of rooibos tea extracts as an objective colour quality parameter," International Journal of Food Science and Technology, vol. 30, pp. 783–792, 1995.

58. E. Joubert, M. Manley, and M. Botha, "Evaluation of spectrophotometric methods for screening of green rooibos (Aspalathus linearis) and green honeybush (Cyclopia genistoides) extracts for high levels of bio-active compounds," Phytochemical Analysis, vol. 19, no. 2, pp. 169–178, 2008.

59. E. Joubert and R. Müller, "A small-scale rotary fermentation unit for rooibos tea," International Journal of Food Science and Technology, vol. 32, no. 2, pp. 135–139, 1997.
60. E. Joubert, F. Otto, S. Grüner, and B. Weinreich, "Reversed-phase HPLC determination of mangiferin, isomangiferin and hesperidin in Cyclopiaand the effect of harvesting date on the phenolic composition of C. genistoides," European Food Research and Technology, vol. 216, no. 3, pp. 270–273, 2003.
61. E. Joubert, E. S. Richards, J. D. Van Der Merwe, D. De Beer, M. Manley, and W. C. A. Gelderblom, "Effect of species variation and processing on phenolic composition and in vitro antioxidant activity of aqueous extracts of Cyclopia spp. (Honeybush tea)," Journal of Agricultural and Food Chemistry, vol. 56, no. 3, pp. 954–963, 2008.
62. E. Joubert and H. Schulz, "Production and quality aspects of rooibos tea and related products. A review," Journal of Applied Botany and Food Quality, vol. 80, no. 2, pp. 138–144, 2006.
63. E. Joubert, J. D. Van der Merwe, W. C. A. Gelderblom, and M. Manley, "Implication of CYP450 stabilization in the evaluation of in vitroantimutagenicity of the herbal teas, Cyclopia spp. (honeybush) andAspalathus linearis (rooibos) and selected polyphenols," in Proceedings of the Polyphenols Communications 2006: Abstracts of 23rd International Conference on Polyphenols, pp. 505–506, Manitoba, Canada, 2006.
64. E. Joubert, P. Winterton, T. J. Britz, and D. Ferreira, "Superoxide anion and α, α-diphenyl-β-picrylhydrazyl radical scavenging capacity of rooibos (Aspalathus linearis) aqueous extracts, crude phenolic fractions, tannin and flavonoids," Food Research International, vol. 37, no. 2, pp. 133–138, 2004.
65. E. Joubert, P. Winterton, T. J. Britz, and W. C. A. Gelderblom, "Antioxidant and pro-oxidant activities of aqueous extracts and crude polyphenolic fractions of rooibos (Aspalathus linearis)," Journal of Agricultural and Food Chemistry, vol. 53, no. 26, pp. 10260–10267, 2005.
66. P. Mose Larsen, S. J. Fey, J. Louw, and L. Joubert, "An anti-diabetic extract of Rooibos," PCT Application PCT/EP2008/052861 (WO, 2008/110551 A1), 2008.

67. H.-K. Na, K. S. Mossanda, J.-Y. Lee, and Y.-J. Surh, "Inhibition of phorbol ester-induced COX-2 expression by some edible African plants," BioFactors, vol. 21, no. 1–4, pp. 149–153, 2004.
68. F. Bruno and W. Dimpfel, "Aspalathin-like dihydrochalcone, extracts from unfermented rooibos and process for preparation," PCT Patent Application PCT/EP2008/007279 (WO/2009/052895), 2009.
69. B. Brinkhaus, M. Lindner, D. Schuppan, and E. G. Hahn, "Chemical, pharmacological and clinical profile of the East Asian medical plantCentella asiatica," Phytomedicine, vol. 7, no. 5, pp. 427–448, 2000.
70. W.-J. Kim, J. Kim, B. Veriansyah et al., "Extraction of bioactive components from Centella asiatica using subcritical water," Journal of Supercritical Fluids, vol. 48, no. 3, pp. 211–216, 2009.
71. A. Shukla, A. M. Rasik, G. K. Jain, R. Shankar, D. K. Kulshrestha, and B. N. Dhawan, "In vitro and in vivo wound healing activity of asiaticoside isolated from Centella asiatica," Journal of Ethnopharmacology, vol. 65, no. 1, pp. 1–11, 1999.
72. J. Lee, E. Jung, Y. Kim et al., "Asiaticoside induces human collagen I synthesis through TGFβ receptor I kinase (T RI kinase)-independent Smad signaling," Planta Medica, vol. 72, no. 4, pp. 324–328, 2006.
73. M. T. Thomas, R. Kurup, A. J. Johnson et al., "Elite genotypes/chemotypes, with high contents of madecassoside and asiaticoside, from sixty accessions of Centella asiatica of south India and the Andaman Islands: for cultivation and utility in cosmetic and herbal drug applications," Industrial Crops and Products, vol. 32, no. 3, pp. 545–550, 2010.
74. H. S. Long, M. A. Stander, and B. E. Van Wyk, "Notes on the occurrence and significance of triterpenoids (asiaticoside and related compounds) and caffeoylquinic acids in Centella species," South African Journal of Botany, vol. 82, pp. 53–59, 2012.
75. G. I. Stafford, M. E. Pedersen, J. van Staden, and A. K. Jäger, "Review on plants with CNS-effects used in traditional South African medicine against mental diseases," Journal of Ethnopharmacology, vol. 119, no. 3, pp. 513–537, 2008.

76. A. Gurib-Fakim, M. Sewraj, J. Gueho, and E. Dulloo, "Medicalethnobotany of some weeds of Mauritius and Rodrigues,"Journal of Ethnopharmacology, vol. 39, no. 3, pp. 175–185, 1993.
77. Y. K. Gupta, M. H. Veerendra Kumar, and A. K. Srivastava, "Effect ofCentella asiatica on pentylenetetrazole-induced kindling, cognition and oxidative stress in rats," Pharmacology Biochemistry and Behavior, vol. 74, no. 3, pp. 579–585, 2003.
78. T. K. Chatterjee, A. Chakraborty, M. Pathak, and G. C. Sengupta, "Effects of plant extract Centella asiatica (Linn.) on cold restraint stress ulcer in rats," Indian Journal of Experimental Biology, vol. 30, no. 10, pp. 889–891, 1992.
79. M. Ramanathan, S. Sivakumar, P. R. Anandvijayakumar, C. Saravanababu, and P. R. Pandian, "Neuroprotective evaluation of standardized extract of centella asciatica in monosodium glutamate treated rats," Indian Journal of Experimental Biology, vol. 45, no. 5, pp. 425–431, 2007.
80. D. M. Pereira, J. Faria, L. Gaspar et al., "Exploiting Catharanthus roseusroots: source of antioxidants," Food Chemistry, vol. 121, no. 1, pp. 56–61, 2010.
81. A. Gurib-Fakim, J. Gueho, and M. D. Sewraj, Plantes Medicinales de Maurice, Editions Le Printemps, Rose Hill, Mauritius, 1995.
82. F. Ferreres, D. M. Pereira, P. Valentão, P. B. Andrade, R. M. Seabra, and M. Sottomayor, "New phenolic compounds and antioxidant potential ofCatharanthus roseus," Journal of Agricultural and Food Chemistry, vol. 56, no. 21, pp. 9967–9974, 2008.
83. D. M. Pereira, F. Ferreres, J. Oliveira, P. Valentão, P. B. Andrade, and M. Sottomayor, "Targeted metabolite analysis of Catharanthus roseus and its biological potential," Food and Chemical Toxicology, vol. 47, no. 6, pp. 1349–1354, 2009.
84. R. van der Heijden, D. I. Jacobs, W. Snoeijer, D. Hallard, and R. Verpoorte, "The Catharanthus alkaloids: pharmacognosy and biotechnology," Current Medicinal Chemistry, vol. 11, no. 5, pp. 607–628, 2004.
85. J. K. Grover, S. Yadav, and V. Vats, "Medicinal plants of India with anti-diabetic potential," Journal of Ethnopharmacology, vol. 81, no. 1, pp. 81–100, 2002.

86. J. Bowie, "Sketches of the botany of South Africa," South African Quarterly Journal, pp. 27–36, 1830.
87. J. M. Watt and M. G. Breyer-Brandwijk, Medicinal and Poisonous Plants of Southern and Eastern Africa, E&S Livingstone, Edinburgh, UK, 2nd edition, 1962.
88. R. Marloth, The Flora of South Africa with Synoptical Tables of the Genera of the Higher Plants, Darter Bros & Co, Cape Town, South Africa, 1925.
89. B. Rood, Uit Die Veldapteek, Tafelberg-Uitgewers Bpk, Cape Town, South Africa, 1994.
90. A. Kokotkiewicz, M. Luczkiewicz, J. Pawlowska et al., "Isolation of xanthone and benzophenone derivatives from Cyclopia genistoides (L.) Vent. (honeybush) and their pro-apoptotic activity on synoviocytes from patients with rheumatoid arthritis," Fitoterapia, vol. 90, pp. 199–208, 2013.
91. M. L. Andersen, E. H. R. Santos, M. D. L. V. Seabra, A. A. B. Da Silva, and S. Tufik, "Evaluation of acute and chronic treatments withHarpagophytum procumbens on Freund's adjuvant-induced arthritis in rats," Journal of Ethnopharmacology, vol. 91, no. 2-3, pp. 325–330, 2004.
92. N. Mncwangi, W. Chen, I. Vermaak, A. M. Viljoen, and N. Gericke, "Devil's Claw-A review of the ethnobotany, Phytochemistry and biological activity of Harpagophytum procumbens," Journal of Ethnopharmacology, vol. 143, pp. 755–771, 2012.
93. S. Chrubasik and P. R. Bradley, "Addendum to the ESCOP monograph on Harpagophytum procumbens (multiple letter)," Phytomedicine, vol. 11, no. 7-8, pp. 691–695, 2004.
94. T. Wegener and N.-P. Lüpke, "Treatment of Patients with Arthrosis of Hip or Knee with an Aqueous Extract of Devil's Claw (Harpagophytum procumbens DC.)," Phytotherapy Research, vol. 17, no. 10, pp. 1165–1172, 2003.
95. E. Ernst and S. Chrubasik, "Phyto-anti-inflammatories: a systemic review of randomized, placebo- controlled, double-blind trials," Rheumatic Disease Clinics of North America, vol. 26, no. 1, pp. 13–27, 2000.
96. P. Wenzel and T. Wegener, "Harpagophytum procumbens—a plant antirheumatic agent," Deutsche Apotheker Zeitung, vol. 135, no. 13, pp. 15–28, 1995

97. N. Ahmad, M. R. Hassan, H. Halder, and K. S. Bennoor, "Effect ofMomordica charantia (Karolla) extracts on fasting and postprandial serum glucose levels in NIDDM patients," Bangladesh Medical Research Council Bulletin, vol. 25, no. 1, pp. 11–13, 1999.
98. Y. Kimura, Y. Minami, T. Tokuda, S. Nakajima, S. Takagi, and G. Funatsu, "Primary structures of N-linked oligosaccharides of momordin-a, a ribosome-inactivating protein from Momordica charantia seeds,"Agricultural and Biological Chemistry, vol. 55, no. 8, pp. 2031–2036, 1991.
99. M. F. Mahomoodally, A. Gurib-Fakim, and A. H. Subratty, "A kinetic model for in vitro intestinal uptake of l-tyrosine and d(+)-glucose across rat everted gut sacs in the presence of Momordica charantia, a medicinal plant used in traditional medicine against diabetes mellitus," Journal of Cell and Molecular Biology, vol. 3, pp. 39–44, 2004.
100. H. Matsuda, Y. Li, T. Murakami, N. Matsumura, J. Yamahara, and M. Yoshikawa, "Antidiabetic principles of natural medicines. III. Structure-related inhibitory activity and action mode of oleanolic acid glycosides on hypoglycemic activity," Chemical and Pharmaceutical Bulletin, vol. 46, no. 9, pp. 1399–1403, 1998.
101. H. Matsuura, C. Asakawa, M. Kurimoto, and J. Mizutani, "α-Glucosidase inhibitor from the seeds of balsam pear (Momordica charantia) and the fruit bodies of Grifola frondosa," Bioscience, Biotechnology and Biochemistry, vol. 66, no. 7, pp. 1576–1578, 2002.
102. P. Khanna, S. C. Jain, A. Panagariya, andV. P. Dixit, "Hypoglycemic activity of polypeptide-p from a plant source," Journal of Natural Products, vol. 44, no. 6, pp. 648–655, 1981.
103. M. S. Akhtar, M. A. Athar, and M. Yaqub, "Effect of Momordica charantia on blood glucose level of normal and alloxan-diabetic rabbits,"Planta Medica, vol. 42, no. 3, pp. 205–212, 1981.
104. D. K. Dubey, A. R. Biswas, J. S. Bapna, and S. C. Pradhan, "Hypoglycaemic and antihyperglycaemic effects of Momordica charantiaseed extracts in albino rats," Fitoterapia, vol. 58, no. 6, pp. 387–390, 1987.
105. E. H. Karunanayake, J. Welihinda, S. R. Sirimanne, and G. Sinnadorai, "Oral hypoglycaemic activity of some medicinal

plants of Sri Lanka,"Journal of Ethnopharmacology, vol. 11, no. 2, pp. 223–231, 1984.

106. A. P. Jayasooriya, M. Sakono, C. Yukizaki, M. Kawano, K. Yamamoto, and N. Fukuda, "Effects of Momordica charantia powder on serum glucose levels and various lipid parameters in rats fed with cholesterol-free and cholesterol- enriched diets," Journal of Ethnopharmacology, vol. 72, no. 1-2, pp. 331–336, 2000.

107. S. Sarkar, M. Pranava, and R. A. Marita, "Demonstration of the hypoglycemic action of Momordica charantia in a validated animal model of diabetes," Pharmacological Research, vol. 33, no. 1, pp. 1–4, 1996.

108. E. Yesilada, I. Gurbuz, and H. J. Shibata, "Momordica charantia: an overview," Journal of Ethnopharmacology, vol. 66, pp. 289–293, 1999.

109. A. Raman and C. Lau, "Antidiabetic properties and phytochemistry ofMomordica charantia L., (Cucurbitaceae)," Phytomedicine, vol. 2, pp. 349–362, 1996.

110. B. A. Shibib, L. A. Khan, and R. Rahman, "Hypoglycaemic activity ofCoccinia indica and Momordica charantia in diabetic rats: depression of the hepatic gluconeogenic enzymes glucose-6-phosphatase and fructose-1,6-bisphosphatase and elevation of both liver and red-cell shunt enzyme glucose-6-phosphate dehydrogenase," Biochemical Journal, vol. 292, no. 1, pp. 267–270, 1993.

111. J. Welihinda, G. Avidson, E. Gylte, B. Hellman, and E. Karlson, "The insulin-releasing activity of the tropical plant Momordica charantia,"Acta Biology Medicinal Germany, vol. 41, pp. 1229–1239, 1981.

112. J. Welihinda and E. H. Karunanayake, "Extra-pancreatic effects ofMomordica charantia in rats," Journal of Ethnopharmacology, vol. 17, no. 3, pp. 247–255, 1986.

113. H. Matsuura, C. Asakawa, M. Kurimoto, and J. Mizutani, "α-Glucosidase inhibitor from the seeds of balsam pear (Momordica charantia) and the fruit bodies of Grifola frondosa," Bioscience, Biotechnology and Biochemistry, vol. 66, no. 7, pp. 1576–1578, 2002.

114. A. H. Subratty, A. Gurib-Fakim, and F. Mahomoodally, "Bitter melon: an exotic vegetable with medicinal values," Nutrition and Food Science, vol. 35, no. 3, pp. 143–147, 2005.
115. M. F. Mahomoodally, A.-G. Fakim, and A. H. Subratty, "Momordica charantia extracts inhibit uptake of monosaccharide and amino acid across rat everted gut sacs in vitro," Biological and Pharmaceutical Bulletin, vol. 27, no. 2, pp. 216–218, 2004.
116. M. F. Mahomoodally, A. Gurib-Fakim, and A. H. Subratty, "Experimental evidence for in vitro fluid transport in the presence of a traditional medicinal fruit extract across rat everted intestinal sacs,"Fundamental and Clinical Pharmacology, vol. 19, no. 1, pp. 87–92, 2005.
117. M. F. Mahomoodally, A. Gurib Fakim, and A. H. Subratty, "Stimulatory effects of Antidesma madagascariense on D-glucose, L-tyrosine, fluid and electrolyte transport across rat everted intestine, comparable to insulin action in vitro," British Journal of Biomedical Science, vol. 63, no. 1, pp. 12–17, 2006.
118. A. Timmer, J. Günther, G. Rücker, E. Motschall, G. Antes, and W. V. Kern, "Pelargonium sidoides extract for acute respiratory tract infections," Cochrane Database of Systematic Reviews, no. 3, Article ID CD006323, 2008.
119. T. B. Agbabiaka, R. Guo, and E. Ernst, "Pelargonium sidoides for acute bronchitis: a systematic review and meta-analysis," Phytomedicine, vol. 15, no. 5, pp. 378–385, 2008.
120. A. Conrad, C. Hansmann, I. Engels, F. D. Daschner, and U. Frank, "Extract of Pelargonium sidoides (EPs) improves phagocytosis, oxidative burst, and intracellular killing of human peripheral blood phagocytes in vitro," Phytomedicine, vol. 14, no. 1, pp. 46–51, 2007.
121. A. Conrad, I. Jung, D. Tioua et al., "Extract of Pelargonium sidoides(EPs) inhibits the interactions of group A-streptococci and host epitheliain vitro," Phytomedicine, vol. 14, no. 1, pp. 52–59, 2007.
122. A. Conrad, H. Kolodziej, and V. Schulz, "Pelargonium sidoides-extract (EPs 7630): registration confirms efficacy and safety," Wiener Medizinische Wochenschrift, vol. 157, no. 13-14, pp. 331–336, 2007.

123. A. Conrad and U. Frank, "Extract of Pelargonium sidoides (EPs 7630) displays anti-infective properties by enhanced phagocytosis and differential modulation of host-bacteria interactions," Planta Medica, vol. 74, no. 6, pp. 682–685, 2008.

124. A. Conrad, D. Bauer, C. Hansmann, I. Engels, and U. Frank, "Extract ofPelargonium sidoides (EPs 7630) improves phagocytosis, oxidative burst, and intracellular killing of human peripheral blood phagocytes in vitro,"Zeitschrift fur Phytotherapie, vol. 29, no. 1, pp. 15–18, 2008.

125. A. Conrad, D. Bauer, I. Jung et al., "Extract of Pelargonium sidoides (EPs 7630) inhibits the interactions of group A-streptococci and host epithelia," Zeitschrift fur Phytotherapie, vol. 29, no. 1, pp. 19–22, 2008.

126. T. Brendler and B.-E. van Wyk, "A historical, scientific and commercial perspective on the medicinal use of Pelargonium sidoides (Geraniaceae),"Journal of Ethnopharmacology, vol. 119, no. 3, pp. 420–433, 2008.

127. O. Kayser, Phenolische Inhaltsstoffe von Pelargonium sidoides DC. und Untersuchungen zur Wirksamkeit der Umcka-Droge (Pelargonium sidoides DC. und Pelargonium reniforme Curt.) [Ph.D. thesis], University of Berlin, 1997.

128. F. B. Lewu, D. S. Grierson, and A. J. Afolayan, "Extracts fromPelargonium sidoides inhibit the growth of bacteria and fungi,"Pharmaceutical Biology, vol. 44, no. 4, pp. 279–282, 2006.

129. F. Daschner, A. Dorfmüller, I. Engels, and U. Frank, "Untersuchungen zum antibakteriellen Wirkmechanismus von EPs7630," Phytopharmaka und Phytotherapie, Abstract, vol. 15, 2004.

130. N. Wittschier, G. Faller, and A. Hensel, "An extract of Pelargonium sidoides (EPs 7630) inhibits in situ adhesion of Helicobacter pylori to human stomach," Phytomedicine, vol. 14, no. 4, pp. 285–288, 2007.

131. W. Beil and P. Kilian, "EPs, an extract from Pelargonium sidoides roots inhibits adherence of Helicobacter pylori to gastric epithelial cells,"Phytomedicine, vol. 14, no. 1, pp. 5–8, 2007.

132. N. Wittschier, C. Lengsfeld, S. Vorthems et al., "Large molecules as anti-adhesive compounds against pathogens," Journal of Pharmacy and Pharmacology, vol. 59, no. 6, pp. 777–786, 2007.
133. P. Taylor, S. Maalim, and S. Coleman, "The strange story of umckaloabo," Pharmaceutical Journal, vol. 275, no. 7381, pp. 790–792, 2005
134. T. Gödecke, M. Kaloga, and H. Kolodziej, "A phenol glucoside, uncommon coumarins and flavonoids from Pelargonium sidoides DC,"Zeitschrift für Naturforschung, vol. 60, no. 6, pp. 677–682, 2005.
135. S. P. N. Mativandlela, J. J. M. Meyer, A. A. Hussein, and N. Lall, "Antitubercular activity of compounds isolated from Pelargonium sidoides," Pharmaceutical Biology, vol. 45, no. 8, pp. 645–650, 2007.
136. O. Kayser, K. N. Masihi, and A. F. Kiderlen, "Natural products and synthetic compounds as immunomodulators," Expert Review of Anti-Infective Therapy, vol. 1, no. 2, pp. 319–335, 2003.
137. H. Kolodziej, "Traditionally used Pelargonium species: chemistry and biological activity of umckaloabo extracts and their constituents,"Current Topics in Phytochemistry, vol. 3, pp. 77–93, 2000.
138. H. Kolodziej, "Fascinating metabolic pools of Pelargonium sidoides andPelargonium reniforme, traditional and phytomedicinal sources of the herbal medicine Umckaloabo," Phytomedicine, vol. 14, no. 1, pp. 9–17, 2007.
139. E. Koch, H. Hauer, and K. H. Stumpf, Use of Pelargonium sidoides andPelargonium reniforme root extracts, World Patent WO2006002837 (European Patent EP1651244, United States Patent 20060263448), 2006.
140. W. R. Sawadogo, M. Schumacher, M. Teiten, M. Dicato, and M. Diederich, "Traditional West African pharmacopeia, plants and derived compounds for cancer therapy," Biochemical Pharmacology, vol. 84, pp. 1225–1240, 2012.
141. R. Gautam, A. Saklani, and S. M. Jachak, "Indian medicinal plants as a source of antimycobacterial agents," Journal of Ethnopharmacology, vol. 110, no. 2, pp. 200–234, 2007.

142. O. M. J. Kasilo and J. M. Trapsida, "Regulation of traditional medicine in the WHO African region," The African Health Monitor, vol. 13, pp. 25–31, 2010.

Citations

CHAPTER 1

Anwar Ali Shad, Shabir Ahmad, Riaz Ullah, et al., "Phytochemical and Biological Activities of Four Wild Medicinal Plants," The Scientific World Journal, vol. 2014, Article ID 857363, 7 pages, 2014, doi:10.1155/2014/857363.

CHAPTER 2

Sabita N. Saldanha and Trygve O. Tollefsbol, "The Role of Nutraceuticals in Chemoprevention and Chemotherapy and Their Clinical Outcomes," Journal of Oncology, vol. 2012, Article ID 192464, 23 pages, 2012, doi:10.1155/2012/192464.

CHAPTER 3

Harvey Babich, Alyssa G. Schuck, Jeffrey H. Weisburg, and Harriet L. Zuckerbraun, "Research Strategies in the Study of the Pro-Oxidant Nature of Polyphenol Nutraceuticals," Journal of Toxicology, vol. 2011, Article ID 467305, 12 pages, 2011. doi:10.1155/2011/467305.

CHAPTER 4

Yuqing Dong, Sujay Guha, Xiaoping Sun, Min Cao, Xiaoxia Wang, and Sige Zou, "Nutraceutical Interventions for Promoting Healthy Aging in Invertebrate Models," Oxidative Medicine and Cellular Longevity, vol. 2012, Article ID 718491, 10 pages, 2012. doi:10.1155/2012/718491.

CHAPTER 5

Naima Saeed, Muhammad R Khan, and Maria Shabbir, Antioxidant activity, total phenolic and total flavonoid contents of whole plant extracts Torilis leptophylla L, doi: 10.1186/1472-6882-12-221.

CHAPTER 6

Sharifa Sultana Dipti, Christine Bergman, Siti Dewi Indrasari, Theja Herath, Robert Hall, Hueihong Lee, Fatemeh Habibi, Priscila Zaczuk Bassinello, Eduardo Graterol, Julie P Ferraz, and Melissa Fitzgerald, The potential of rice to offer solutions for malnutrition and chronic diseases, doi:10.1186/1939-8433-5-16.

CHAPTER 7

M. Fawzi Mahomoodally, "Traditional Medicines in Africa: An Appraisal of Ten Potent African Medicinal Plants," Evidence-Based Complementary and Alternative Medicine, vol. 2013, Article ID 617459, 14 pages, 2013. doi:10.1155/2013/617459.

Index